Mysteries of the Marfa Lights Revealed

A Guide to the History, Mystery, Science, and Viewing of the Marfa Lights

ROBERT WAGERS
AND
JUDY WAGERS

RJ Books Unique

Black and White Edition

Published by R & J Books Unique
Richardson, Texas
www.marfalightsmysteries.com

Cover Photographs
Front: *Mirage on US90 five miles east of Marfa, Texas* by Robert Wagers
Back Top: *Color Rendition of Figure 12–3.*
Back Bottom: *Spectra of The Moon and The Sun* by Robert Wagers

Wagers, Robert and Wagers, Judy
Included glossary and bibliographic sources

ISBN: 0988827123
ISBN-13: 978-0-9888271-2-7

Acknowledgements

WE ARE GRATEFUL FOR THE ASSISTANCE provided by staff members from several organizations during our historical research efforts. Many documents had to be tracked down to their original, hardcopy form. Some of these documents were found at the Presidio County Recorder's Office in Marfa. Others were located in the library archives of campuses of the University of Texas at Arlington, at Dallas, and at El Paso. Southern Methodist University in Dallas and the Archives of the Big Bend at Sul Ross State University in Alpine also held original sources. Jerri Garza at Sul Ross State University was especially helpful — not only in finding items we were seeking, but in suggesting related avenues of search.

Discovering the paths followed by the earliest roads in the Marfa area was no easy task. The United States Geological Survey (USGS) was most helpful in providing maps prior to 1900. After the turn of the century, when automobile traffic was in its infancy, most of the roads followed the original horse and buggy trails shown on the USGS maps. It wasn't until the mid 1930s that there was any effort to improve the regional "highway system". Chris Weber and Becky Lucas of the Texas Department of Transportation were invaluable in helping us identify the old routes which no longer exist, and we gratefully acknowledge their assistance.

—§—

This book is also available in full color print and digital editions.

Mysteries of the Marfa Lights Revealed – Full Color Edition
ISBN: 978-0-9888271-1-0

Mysteries of the Marfa Lights Revealed
ISBN: 978-0-9888271-0-3 (Digital Edition for Kindle and Nook)

Published by

R & J Books Unique
Richardson, Texas
www.marfalightsmysteries.com

Contents

Preface

THE AUTHORS BECAME AWARE OF THE MARFA LIGHTS in 2005 and 2006 when visiting the Big Bend region of Texas. Readily available information in the motels, businesses, and at the Marfa Lights Viewing Center indicated that the lights had been seen for over a century. Yet, there was no agreed upon explanation for their source. There were numerous theories, but none of them was accepted as *THE explanation*. "How could this be possible?" we wondered.

Curious as to the reasons for the mystery, we began examining the various ideas put forth as explanations for the strange lights. While there were many scientific-sounding theories, none of them had a quantitative basis. In fact, in nearly all cases, the theories advanced did not even invoke plausible physics.

It wasn't difficult to eliminate most of the conjectures. But those remaining required some serious efforts in order to assess their merits. Before long, our casual inquiry had turned into a major investigation. As we began to have success, we realized that other people would be interested in our findings too. Thus, this book began. Our goal was to explain the origins of the mystery lights and make it possible for the reader to have a mystery-light-viewing experience that was also understandable.

We established six objectives for the book. Everything begins with the first objective — explain the physical basis of the mystery lights. This challenge is actually multifaceted. To be sure, it requires discovering the sources of the light. But, more than that, it requires an explanation for what might make the light have a mysterious appearance or behavior.

Several chapters are devoted to an examination of a host of possible light sources ranging from campfires to the moon. This is followed by in-depth analyses of the role played by the distortion of light in making the appearances seem mysterious. These results are applied to a collection of eye-witness reports as a means of validating the analyses and illustrating the character of the lights to the reader.

The second objective was to define the prerequisites for repeatability of any of the light sightings. Some lights can be seen every night; others seem to appear randomly in both place and time. Discovering what controls repeatability, provides a basis for future opportunities for others to see the lights. This objective was met as a corollary to the first objective because if you know what the source of light is and what controls its appearance, you also know what enables or prevents the mysterious light.

After discovering what controls repeatability, we wanted to find locations that afford repeatable sightings. The Marfa Lights Viewing Center is certainly the best known site for repeatable viewing of the mystery lights. But, were there more sites in the Marfa area? There have been some in the past. Are they still accessible? We set about to answer this question.

Another very important objective of the book was to set the mystery lights in the context of the history of the Big Bend. Stories of mysterious lights in the Big Bend area have been reported for more than a century. How do those early reports relate to today's lights? What could have caused those early mysterious lights? They certainly were not car lights. But what *were* they?

To make our explanations as concrete as possible and bring the story to life for the reader, we sought photographic illustrations of light behavior that were repeatable and accessible to the public. One such repeatable phenomenon is the mirage on the front cover of this book.

Lastly, we wanted to put forth a guide that could be followed to see the history and natural phenomena of the mystery lights in the Marfa region of the Big Bend. This we accomplished by embedding location information for relevant sites throughout the book.

Many of the concepts and arrangements of features discussed in this book are difficult to communicate in words alone. But, just a glance at some of the 32 illustrations and eight maps reveals an entire story by itself, and leaves a lasting image in the mind of the reader. Ultimately, this was the authors' objective.

Robert Wagers
Judy Wagers
January 25, 2013

—§—

1 — Introduction

EVERYONE LOVES A MYSTERY. And, it is even better if the explanation invokes mystical and supernatural powers. The Greeks had their all-powerful Zeus hurling lightning bolts from Mount Olympus — punctuating them with resounding claps of thunder. No one dared doubt his power over the universe with such a display. The attraction to supernatural powers continues today.

Think back several decades. We had Superman endowed with powers not found on this earth. We read his comic books, watched him on television, and flocked to the movies to watch him leaping tall buildings with a single bound — vulnerable only to Kryptonite. Then came *Star Trek,* with *warp power, phasors,* and *dilithium crystals,* searching out evil from the far corners of the federated universe. Still more recently *Star Wars* brought us *light sabres*, and the supernatural powers of *Jedi warriors.*

Not only do we love heroes with supernatural powers — we accept their extraordinary abilities — and don't even ask how! The fact that we do not question *how* has characterized our acceptance of supernatural explanations from the beginning of recorded history. Or, maybe those who asked questions were eliminated — struck down by Zeus' lightning bolt.

The penchant to explain mysterious observations through reference to equally mysterious causes enjoys the same popularity today as it did in yesteryear. Read the literature about the Marfa lights, and you will discover their existence attributed to causes such as *cosmic rays, telluric currents*, and *ball lightning*. These are exotic, extra worldly causes like *warp power* — all put forth without plausible explanation.

But what of these causes? Could they really be responsible for the mysterious lights at Marfa?

Let's examine them in turn. *Cosmic rays* sound extraterrestrial and conjure up images of Zeus throwing light at the earth. Unfortunately, cosmic rays are less mysterious than their name suggests. They fall on the earth all the time — *and* everywhere. Why would they cause Marfa's mystery lights and not lights elsewhere? No one can explain this exclusivity.

Telluric currents sound as mysterious and supernatural as *dilithium crystals*, so they have natural appeal. But *telluric* just means that it pertains to the earth. If you get rid of the fancy word and say "earth currents" is the appeal the same? Probably not. By the way, electrical earth currents are not exclusive to the Marfa plain either; they are everywhere. So, if earth currents are responsible for Marfa's mysterious lights, they should also cause mysterious lights elsewhere on the planet. Why just Marfa?

Then, there is *ball lightning*. It is popularly offered as an explanation. There is only one problem — no one knows what ball lightning is. It has never been observed in a scientific setting. Thus, it has never been measured and has no basis for its composition or physical properties. The label *ball lightning* seems to have been coined based on reports by people who saw spots of light following a lightning strike. The word *ball* stems from the fact that the spots of light were round, and the word *lightning* was added based on the temporal proximity to the stroke of lightning in the atmosphere.

But seeing spots after a stroke of lightning does not mean that any mystical or supernatural process occurred. Some people just see spots. So, we can find no credible explanation for the mystery of Marfa's lights in ball lightning either. We need to dig deeper for our explanation.

Legend has it that the first sighting of mysterious lights on the Marfa plain occurred in 1883. As it happens, 1883 was also the year in which Lord Kelvin, one of the world's greatest scientists, set forth these thoughts about the minimum requirements for a science —

> *I often say that when you can measure what you are speaking about, and express it in numbers, you know something about it; but when you cannot measure it, when you cannot express it in numbers, your knowledge is of a meagre and unsatisfactory kind; it may be the beginning of knowledge, but you have scarcely, in*

> *your thoughts, advanced to the stage of science, whatever the matter may be.*
>
> — Lord Kelvin, May 3, 1883[1]

Well, if we follow Lord Kelvin, then all the mystical and supernatural examples above are clearly *not* science. Neither are the explanations in the literature that treat the Marfa lights as manifestations of other, even more obscure, phenomena.

But does that mean that the truth about the Marfa lights is as boring as some of the math and physics classes that many of us have had to endure? Not at all!

Nature offers some very interesting phenomena. If we focus on providing explanations of the phenomena and how they enable unexpected light behavior — instead of focusing on the math and physics, the investigation of Marfa lights is a very fascinating story.

In this book, we will show how historic observations are consistent with known physical processes. Our analyses will demonstrate how the processes could have caused the observed light phenomena. While we can explain light phenomena with physics, we cannot ***prove*** what happened historically — nobody can. Those events are gone; all that remains are reports from the observers.

Some sightings that repeat regularly are easily explained. For example, lights of vehicles approaching Marfa from the south on US67 are as predictable as the sunset, and you can easily go to the location of the light sources and see the vehicles.

But other sightings are not as easily explained. Predicting future sightings of mysterious lights around the Marfa plain is difficult because truly mysterious lights repeat so infrequently. They require a source of light to be in an appropriate location, the terrain and atmospheric conditions to allow the light to be seen, and the observer to be in just the right spot at just the right time to see the phenomenon. These are demanding constraints — especially today when most of the Marfa plain is private property and field trips are impossible. Add to this, the fact that there are no analyses of the role the atmosphere played in the mysteries of the Marfa lights and no forecasts for where to go to observe these rare phenomena, and you have a set of strong barriers to a full understanding of these light displays.

Yet, there are compelling stories of mysterious light sightings that lend themselves to rigorously consistent analysis. In turn, these analyses might lead to predictions of future mysterious light occurrences. However, when it comes to predicting a rare repetition of a mysterious light, we will have to caveat the forecast with the requirement for unique circumstances to prevail, as we will explain in the succeeding chapters.

Much of the renown of Marfa derives from its mysterious lights. These lights have a history that is larger than the popular mythology. And that history does not stand alone. It is intertwined with the opening of the West to commerce and travel. The story begins in 1850 with the exploration of west Texas and the subsequent diffusion into the Big Bend region of settlers, cattle, soldiers, ***and their attendant lights***. It continues with the legend of the first mysterious light sighting on the Marfa plain in 1883 and works up to the present considering all the lights employed over the years.

Because so many factors are relevant to the appearance of a "mysterious" light, we will look at topography, meteorology, and the phenomena of lights that span the gamut from fires, matches, and stagecoach lights all the way out to moonbeams. We will examine how light travels from one point to another and the influence of atmospheric layers on the path light takes. Actual light values measured from light sources will be combined with the known visual ability of the human eye to define limits of visibility.

Diverse facts about the Marfa plain and the special conditions that led to its famous mysterious lights will be presented. Our findings will be illustrated with examples from the Marfa area of both distorted daytime images — mirages, and a representative collection of explanations for mysterious nighttime lights that have been reported by eye witnesses in recent years.

Being true to Lord Kelvin's dictate for science, all our explanations will rely on physics that can be confirmed.

Read on for a journey of discovery as we unveil the secrets of the Marfa lights.

—§—

1— Sources

[1] Sir William Thomson [Lord Kelvin], "Electrical Units of Measurements." Lecture delivered at the Institution of Civil Engineers in London on May 3, 1883. Reprinted in Sir William Thomson, *Popular Lectures and Addresses, Volume 1: Constitution of Matter* (London: MacMillan and Company, 1891), 80.

2 — The Marfa Plain

KNOWN FOR ITS "MYSTERIOUS" LIGHTS and the Marfa Lights Festival each Labor Day weekend, Marfa, Texas is a melting pot of cultures — a mix of cowboys, artists, and ordinary folks in an environment where the frontier past meets the present. From its restored 19th century Presidio County Courthouse and Paisano Hotel, to its eclectic El Cosmico and modern art galleries, Marfa is distinct among west Texas towns.

In the middle of the Big Bend region of Texas, Marfa is the county seat of Presidio County — a role it obtained in 1885 after the Galveston, Harrisburg, and San Antonio Railroad was extended through what would become Marfa, bypassing Fort Davis. It is also the town that gave the Marfa plain its name.

How did a town in west Texas come to be called *Marfa*? The simple answer is — the town, which was established in 1883, adopted the name of the railroad watering stop that preceded it. So, how did a railroad stop on the banks of Alamito Creek in west Texas come to be called *Marfa*? It certainly wasn't named for its Mexican or Indian cultures and languages.

The earliest reference to the site being called *Marfa* came from a telegram written on January 21, 1882, and published the next day in *The Galveston Daily News.*[1][1] While the article clearly established the date for the naming of this railroad stop, it did not describe how the name *Marfa* was chosen.

Subsequently, on December 17, 1882, the *The Galveston Daily News* attributed the railroad station's name to a Jules Verne character — Marfa Strogoff — in the play "Michael Strogoff"[2] — which was popular at the time. Other sources go further and say that Hannah Strobridge, the wife of the railroad's chief engineer, was reading the novel *Michael Strogoff* when the tracks reached Marfa, and she proposed the name for this stop.[3],[4] Whether Hannah Strobridge was reading the play or the novel, it is clear from *The Galveston Daily News* article[2] that Marfa was named after the Jules Verne character Marfa Strogoff.

There have been other theories for the naming of Marfa, such as the one on the plaque at the Marfa Lights Viewing Cen-

---

1 The first reference the authors have found to Marfa was a January 21st, 1882, telegram on the progress of railroad building filed by an El Paso reporter for *The Galveston Daily News*. Published the following day, the report referenced the railroad's chief engineer, James Henry Strobridge, as saying that the railroad was complete to Marfa and that crews were busy digging wells at Marfa and other stops west of there.[1]

ter which states that the name came from one of the characters in Dostoyevsky's *The Brothers Karamazov*. But, a newspaper account written contemporaneously with the actual naming is more likely to be correct than a supposition penned close to a century later.

Many people received their first glimpse of Marfa and Presidio County in movies. The 1955 film classic *Giant* with Rock Hudson, Elizabeth Taylor, and James Dean was shot on a ranch near Marfa. And, in 2006, *No Country for Old Men* and *There Will Be Blood* were filmed near Marfa. Forty miles further south on US67, Shafter was the setting for the opening scenes of the *Andromeda Strain*. And, although not filmed on location, the 1959 movie *Rio Bravo* starring John Wayne, Dean Martin, Ricky Nelson, Walter Brennan, and Angie Dickinson was set in the town of Presidio. Today, there are functioning movie sets at local ranches near Alpine and Marfa, and in the Big Bend State Park.

But, there is more to Marfa than light festivals and movies. One of the first things you'll notice as you travel through the region around Marfa is the vast expanse of wide, flat plains interrupted at their edges by mountains that rise a thousand feet or more above the floor of the plains. No where is this more evident than driving into Marfa from the north on US90. A breathtaking panorama unfolds as the road steadily rises until it reaches nearly 5,000 feet above sea level at Marfa. With the elevation, the flatness of the plains, and the expansiveness of the plains as the backdrop, this region of the Big Bend is ideal for sightings of strange lights.

It is also the ideal setting for those interested in geology. The area around Marfa, depicted on the map in Figure 2–1, was shaped by volcanism tens of millions of years ago.

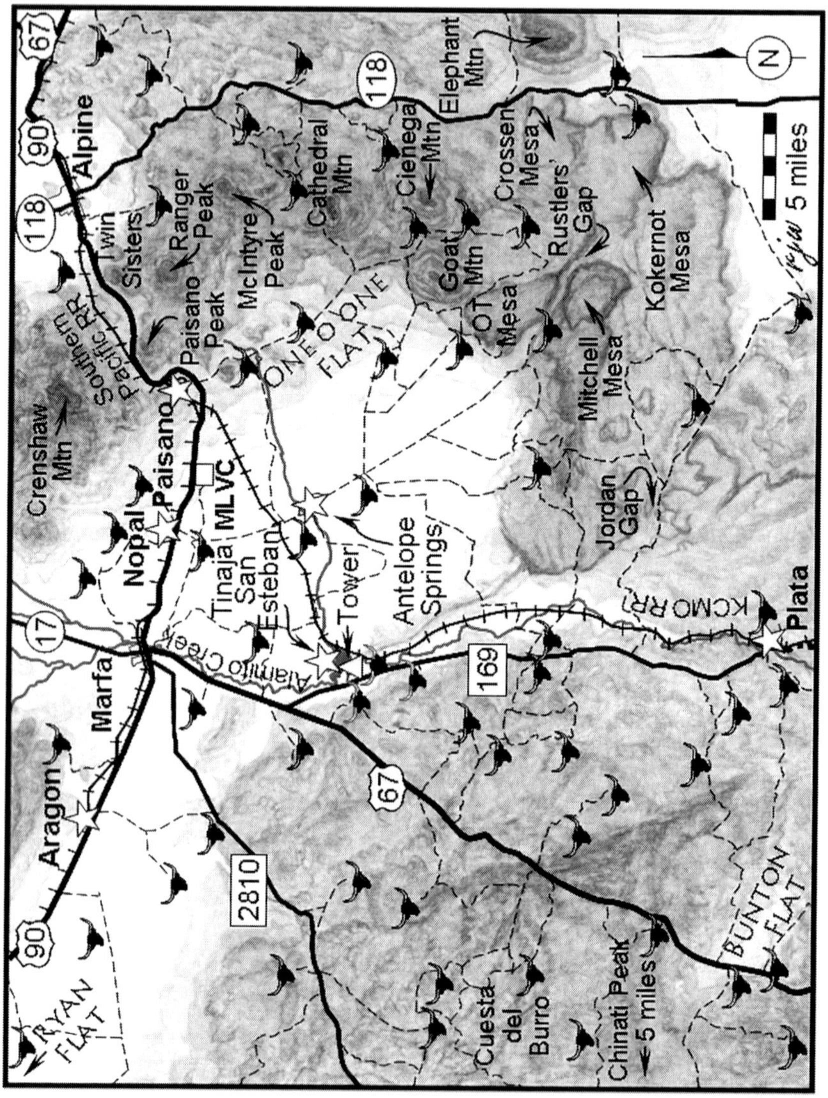

Figure 2–1. Ranches, ranch roads, mountains, and a few points of interest in the vicinity of the Marfa plain.

Today, evidence of volcanism abounds throughout the region. In the Chisos Mountains of Big Bend National Park 75 miles south of Alpine, visitors are surrounded by geological history. They can climb the Chisos' volcanic highland trails to view spectacular natural vistas and camp in the caldera of the mountain's extinct volcano.

Closer to Marfa, on the northeastern portion of the map in Figure 2–1, the drive between Marfa and Alpine along US90[2] takes travelers through Paisano Pass and around Paisano Peak (summit: 6,091 feet).

From the dikes and jagged cliffs to Paisano Peak, everywhere you look there is evidence of ancient lava flows and volcanic activity. Especially impressive are the magma dikes that look like the dorsal fin of a giant dinosaur or sea monster sticking up through the earth's crust. The eruptions which produced the dikes and Paisano Peak occurred over a million-year period around 36 million years ago.

All this geologic activity millions of years ago ultimately created the plains and uplifted them a mile above sea level. The result is an ideal stage for the display of unusual — and some might say *mysterious* — light phenomena. But, before we discuss lights, we need to examine the features of the terrain around Marfa that have figured in the history of the mysterious lights.

The Marfa plain lies at the center of Figure 2–1 surrounded by hills and mountains that are shaded in dark gray. Extending nearly 20 miles in some directions, the plain slopes downward from a high elevation of about 5,000 feet at the northeast corner to about 4,500 feet at the southwest corner where it meets the Alamito Creek.

---

[2] US67 and US90 share the same roadbed between Marfa and Alpine. Outside of Alpine, US67 branches off to the north, while US90 takes an easterly route. For simplicity, we will refer to the stretch of highway between Marfa and Alpine as US90.

With headwaters rooted in the foothills of the Davis Mountains north of Marfa, Alamito Creek runs south 60 miles from Marfa to the Rio Grande, joining the muddy river about seven miles downstream from Presidio. This waterway forms the western boundary of the Marfa plain and was vital to habitation and transportation in the region throughout history.

Three expanses on Figure 2–1 have been designated as *flats* (see *Glossary*) by the United States Geological Survey (USGS). They are the Ryan Flat to the west, the Bunton Flat to the southwest, and the One O One Flat to the east. If you've ever been to this region of Texas, you'd probably say the entire non-mountainous area is flat. But these three regions are flatter than the rest. The One O One Flat is part of the central area called the Marfa plain and figures prominently in our discussion of nocturnal lights in *Chapter 10 – Lights on the One O One Flat*.

North of US90, the terrain rises steadily and is punctuated with peaks reaching up to 6,000 feet. Toward the upper right side of the map, the peaks become more organized with Crenshaw Mountain (5,965 feet) dominating. The range of peaks is interrupted on the east side by Paisano Pass, which has provided a route in and out of the Marfa plain for centuries. In addition to US90, the Southern Pacific Railroad[3] travels through Paisano Pass on its way between Marfa and Alpine. Currently, at 5,093 feet, Paisano Pass is the highest elevation achieved by the Southern Pacific anywhere between California and Galveston, Texas.

Railroads played a dominant role in opening up the Marfa plain to settlement and commerce as we will see in *Chapter 4 –*

[3] The railroad is currently called the *Union Pacific*.

Bringing Lights to Marfa. The tracks of the Southern Pacific Railroad cross the northern Marfa plain from west to east, while those of the Kansas City, Mexico, and Orient Railway (KCMO) cut diagonally across the plain from Paisano Pass to Alamito Creek before turning south to the Mexican border at Presidio. For much of its distance, the Southern Pacific Railroad tracks run parallel to US90 and are shown as a merged highway–railroad symbol in Figure 2–1. When the highway and tracks diverge, individual symbols have been used.

Although the train tracks through the area have deviated little from their original routes, the names of railroads have come and gone with each new owner. The KCMO is no exception. Initiated in 1900 as a link to bring goods from Mexico to the Midwest, its tracks from Kansas City didn't reach Alpine until 1913. Presidio was connected in 1930, and the link to the Gulf of California at Topolobampo, Mexico was added in 1961. The railroad was never a success. Its ownership and name have changed repeatedly over time, and could change again tomorrow. For our purposes, we'll stick with its original name, KCMO.

Similarly, the original railroad that came through Marfa in 1882 was not the Southern Pacific. It was called the Galveston, Harrisburg, and San Antonio (GHSA). This railroad has also repeatedly changed names and ownership over time, but the railroad is certainly most identified with the Southern Pacific name. So, unless there is a specific reason to use a different name for this railroad, we'll refer to it as the *Southern Pacific* in this book.

Railroads of the late 19th and early 20th centuries needed water for their steam engines. When the forerunner of the Southern Pacific Railroad — the GHSA — was extended through the region, watering stations were established at frequent locations. Aragon, Marfa, Nopal, Paisano, and Alpine (then called *Murphysville*) were all stops. Marfa and Alpine went on to become county seats for Presidio and Brewster counties, respectively. The other locations no longer exist, and are designated as historic places, marked by white stars on Figure 2–1.

Although Aragon didn't survive as a watering stop on the Southern Pacific Railroad, residents remained there until after the Second World War. During the war, a small auxiliary air field south of US90 was used by the U.S. Army as part of their pilot training facilities. Because of a unique first-hand description of strange lights seen from Aragon in 1939, we will discuss this region and its mysterious lights in *Chapter 11 – Aragon ~ Lights Below the Chinatis.*

At the bottom of Figure 2–1, Plata is another historic place designated by a white star. Established in the early 1880s as a ranching community, it became a watering stop on the KCMO Railroad sometime after the railroad was extended from Paisano Pass to Presidio in 1930. Today, a rusted water tank stands vigil over the tracks, and decomposing adobe structures are all that remain of the station. The first person reported to have seen mysterious lights on the Marfa plain was a teenage boy, Robert Reed Ellison. Legend has it that he saw the lights one night in March, 1883, while driving a herd of cattle across the Marfa plain on the way to his father's range near Plata.

The highest peaks adjacent to the Marfa plain exist on its east side. Cathedral Mountain rises to 6,868 feet, and Goat Mountain reaches 6,608 feet at its summit. Further south on the east side of the plain are OT Mesa (5,633 feet) and Mitchell Mesa (5,281 feet). The two mesas are separated by Walnut

Draw. That passage is also called Rustlers' Gap — perhaps indicative of its frontier traffic.

Rustlers' Gap is one of a series of readily visible markers of fractures in the earth's crust at the southeast corner of the Marfa plain. They were created millions of years ago. The largest of these faults, the Walnut Draw Fault, arcs around the northern side of the Marfa plain, comes tangent to the southwestern side of the Paisano volcanic region, and heads straight for the southern side of Goat Mountain. It passes between Goat Mountain and OT Mesa and emerges between Mitchell Mesa and Kokernot Mesa.

As a result of these rifts in the earth's surface, OT Mesa, Mitchell Mesa, and the adjacent mountains to the southwest have dropped down from the elevation they would have had without the rifts. OT Mesa, for example, is about 700 feet lower.[5] The Walnut Draw Fault between Goat Mountain and OT Mesa is visible to travelers on US67 and US90 at many locations around the Marfa plain.

The mesas extending west from Mitchell Mesa slope down in elevation and are interrupted by somewhat of a valley where RM169, Alamito Creek, and the KCMO Railway pass out of the Marfa plain on their way south. Then the terrain rises again to the west of RM169 up the slopes of Cuesta del Burro. Both US67 and RM2810 ascend these heights.

Moving back north along the west side of the Marfa plain, a tower in Figure 2–1 is located on the south of San Esteban Lake and directly adjacent to the east side of RM169. Designated by a white triangle, it is at latitude 30.157 N and longitude

104.041 W.[4] The tower's flashing red light is visible for miles after sunset and figures prominently in the sightings of mysterious lights.

When it comes to mysterious lights, perhaps the most important feature is the Marfa Lights Viewing Center (MLVC) shown as a white rectangle just below US90 near the top-center of Figure 2–1. It is 8.5 miles east of Marfa (15.6 miles from Alpine). The concept for the center was developed by the students of Marfa High School. Land for the center was donated by Clayton and Modesta Williams, and the Texas Department of Transportation built the center which opened in 2001.

Every evening people gather there to view the lights that appear to the southwest just after sunset. From the MLVC, the flashing red tower light is quite distinct, even though it is 12.5 miles away. Its position marks the left-hand side of where a particular group of "mysterious" lights appear. A viewer at the MLVC has only to locate the flashing red light and look slightly to the right to see the lights that we'll be discussing in the *Chapter 6 – "Classic" Marfa Lights.*

Marked by white stars on Figure 2–1, two water sources — Antelope Springs (shown just below the MLVC) and Tinaja de San Esteban[5] to the west — are also of historic importance. These two watering sites (plus a string of others across the Big Bend) made travel possible between Presidio and points north.

The two watering sites on the Marfa plain were not only enablers of local life and travel, they were vital to the freight operators who plied the Chihuahua Trail transporting goods

[4] The tower is owned by the Big Bend Telephone Company.

[5] Tinaja de San Esteban now sits at the bottom of San Esteban Lake.

between Indianola, Texas on the Gulf Coast and Chihuahua, Mexico. We'll be mentioning Tinaja de San Esteban and other watering sites again in *Chapter 4 – Bringing Lights to Marfa*, when we examine the development of roads and routes of travel through the Big Bend region in the 1880s.

The final and most numerous features on the map in Figure 2–1 are the ranches (denoted by black longhorns) and ranch roads (designated by dashed lines). Tucked away on mountain sides and scattered across the plains, these ranches are interconnected by an extensive web of ranch roads. One thing that stands out on the map is that on the Marfa plain and the surrounding mountain sides, there are ranches, ranch roads, ranch-to-market roads, state and U.S. highways, and railroads in just about any direction you look. This network makes for a lot of vehicle traffic. And, as we'll see later, vehicle traffic will figure in many of the sightings of mysterious lights in this book.

With our orientation to the Marfa plain complete, in the next chapter, we will begin our investigation of mysterious lights by examining reports of their sighting. An understanding of these reports will guide us in establishing explanations for the different sources of mysterious lights.

—§—

2 — Sources

[1] "El Paso. Notes on Railroad Building—The Southern Pacific—Mexican-Central—Ice Factory," *The Galveston Daily News*, January 22, 1882, p. 1, col. 8.

[2] "Presidio," *The Galveston Daily News*, December 17, 1882, p. 7, col. 5.

[3] *Southern Pacific Lines, The Galveston, Harrisburg & San Antonio Railway, Brief History of El Paso Division 1881-1925*, compiled by Hugh B. Wilson. (El Paso, TX: [n.p.], 1925), 56-57.

[4] Henry T. Fletcher, "Texas Collection," ed. H. Bailey Carroll, *Southwestern History Quarterly* 48, no. 2 (1944): 295.

[5] William N. McAnulty, "Geology of Cathedral Mountain Quadrangle, Brewster County, Texas," *Bulletin of the Geological Society of America* 66, no.5 (May 1955): 564.

3 — Sightings of Mysterious Lights

FIRST-HAND SIGHTINGS OF STRANGE LIGHTS in the Marfa region have been reported for close to a century. Second-hand reports go back even further. As you might imagine, these reports vary widely in their quality, character, and level of detail. With the availability of the Internet, the volume of material about the mysterious lights on the Marfa plain has exploded.

For several years, we have performed Internet searches for the phrase "Marfa lights". In June of 2006, we got more than 30,000 hits. By June of 2011, the hits had grown to 145,000. While the volume is astounding, the quality is low. The vast majority of the material is simply repeated from other websites.

Some web pages claiming to show *"THE* Marfa lights" include videos of the lights from vehicles on US67. To be sure, those headlights are part of the collection of lights that has come to be known as the *Marfa lights*. But the total mystery is comprised of more sightings than just car lights on US67, and the literature on these other observations is exceedingly thin.

In our review of the literature and Internet searches, we've sought descriptions that have the potential for explanation. In this chapter, we'll review the stories that comprise a significant part of the literature about the Marfa lights and analyze their credibility. We will set aside those that are too lacking in detail to warrant further pursuit and review some of the most intriguing ones in greater detail. At the end of this chapter, we'll tabulate the reports that we'll be examining in subsequent chapters.

As we review the stories of mysterious lights, we'll divide them into three categories — first-hand reports, second-hand reports, and folktales. First-hand reports — regardless of their quality — have a better chance of being true sightings than second-hand reports. The stories of folklore cannot be corroborated but are fun to recount and hear. Like tales told around the campfire, they aren't intended to be verifiable.

The more detailed the description of a sighting is, the better. Ideal reports would include the following elements — (1) Photographs or a video of the light; (2) documentation of what was seen, written soon after the sighting; (3) precise statements about the location of the observer, the direction of the view, and the date and time of the observation; and (4) details about recent weather events and current weather conditions such as temperature, pressure, humidity, wind direction and speed, as well as whether atmospheric particulates (smoke and haze) were present. It should come as no surprise that none of the reported sightings includes all of these elements, and most don't even come close to providing **useful** information.

The earliest report of a mysterious light on the Marfa plain tells of a 16-year-old cowboy, Robert Reed Ellison, who herded cattle through Paisano Pass in March of 1883. He was headed southwest across the Marfa plain on his way to his father's range near Plata on Alamito Creek. (Plata is shown at the very bottom of the map in Figure 2–1.)

Many years later, in 1937, Ellison described this trip in his memoirs. He noted that on the first night that he camped on the Marfa plain "a little while after night came, a full moon rose up" and illuminated the mountains south and southwest of him. He remarked that the mountains seemed to have moved closer to him.[1]

Although he remembered the moonrise and the appearance of the mountains, Ellison made no reference to seeing mysterious lights in his memoirs. That portion of the story came from his family. One of his granddaughters, Rosemary Cox explained —

> *What I remember is that there were lights down there that they couldn't explain. They thought they were Indian fires. At least, that's what I remember him saying.*[2]

So, while the trip through the Marfa plain was documented by Robert Ellison and qualifies as a *first-hand report*, the description of mysterious lights is certainly more distant than a *second-hand report*. The best it could be is third hand — from Robert Reed Ellison to his family to authors of publications and on to readers of their works. Regrettably, Ellison's report of mysterious lights has to be categorized as *folklore*. Yet, the story was repeated until it became a local "fact" and is immortalized on a plaque at the Marfa Lights Viewing Center (MLVC).

However, Robert Reed Ellison's recollection of the mountains moving closer to him is a *first-hand* report that bears further examination. Our investigation of this facet of his story in *Chapter 9 – Larger than Life* will reveal an important underpinning of mysterious light behavior.

Our next story of mysterious lights on the Marfa plain also comes from Cecilia Thompson's book.[3] In it, she recounts that Joe and Anne Humphreys reported seeing mysterious lights in 1885. They were supposed to have related their story to their granddaughter, Anne McCracken Markle. However, documentation for this sighting is not available. Thus, like the Ellison story, it has to be counted as *folklore*.

Perhaps the most accomplished storyteller to add to the lore of the mysterious lights in the Big Bend region was Oscar Waldo (O.W.) Williams. His reports are folklore, to be sure. But they are distinctly different from the rest — they were the first to be written down and published.

Williams was a unique person on the Texas and New Mexico frontier in the 1880s. He had a law degree from Harvard, served as judge in Fort Stockton for a time, knew how to survey land, and had an uncommon inclination to document history and folklore. So, it was his good fortune, on one surveying effort in the Big Bend of Texas, to have a member of his crew who liked to tell stories by the campfire at night.

This was in 1902, and the storyteller was Natividad

Lujan — a Mexican, born in San Carlos,[1] Mexico, around 1830. One of the stories Natividad told was of a great Apache chief named *Alsate*. It seems that Alsate came to his end violently. And, as is often the case in stories like this, his ghost was seen as mysterious lights at a number of places where Alsate had traveled while alive.

Williams took notes about this story and others, and included the notes in letters he mailed back to his children.[4] Later, Williams added more material to that contained in the notes and published two different renditions of the Alsate story in pamphlets.[2][5] Both of these pamphlets were published some time after 1919.[6] Even later (1944), he published the story again in *Frontier Times* as "Alsate — The Last of the Chisos Apaches".[7]

Over the years, others have also associated the ghosts mentioned in Williams' writings with mysterious lights. Perhaps the connection derives from the opening passages of the Alsate story where Natividad Lujan refers to the rays of the rising sun as being the spirit of Alsate. Or, maybe the connection is due to the fact that you can't see a ghost at night unless it is luminous. No one knows for sure.

While Judge Williams did not provide any descriptions of mysterious lights ***he*** had seen in the Big Bend, he did add uniquely to the folklore of the region. His contribution is recognized in the plaque placed at the MLVC by Clayton[3] and

---

[1] San Carlos, Mexico is about 12 miles south of Lajitas, Texas. See Figure 4–1.

[2] The version of "Alsate" in the pamphlet "By the Campfire in the Southwest" was reprinted in Reference [4], pp. 257-264.

[3] The grandson of O.W. Williams, Clayton, was an unsuccessful Texas gubernatorial candidate in 1990.

Modesta Williams with the citation, "Their gift honors O.W. Williams who first wrote of the mysterious lights in the 1880's".

Ruth Bownds Kershner is the source of the oldest documented *first-hand* sightings of Marfa lights that the authors have found. Born in Marfa in 1914, she lived there until 1932. In 1999, she recounted two separate stories about the Marfa lights as part of a larger essay entitled, "Marfa Stories".[8] Although her descriptions didn't include all the elements of a good report that we mentioned above, she covered a few of the more important ones.

In her first story, she told of attending dances at a barn just south of Marfa over a four-year span while she was in high school. She indicated that at those events she saw the Marfa lights off toward the east every evening. In the second story, she described seeing the "classic" Marfa lights.[4] These sightings were made from a location close to the present site of the MLVC and occurred on Friday or Saturday nights between 1928 and 1932. She described the lights as looking like long streaks. Although Ruth didn't specify the direction she was looking, she didn't disagree with the Texas Highway Department's sign indicating where to look.

In both cases, Ruth Kershner's sightings were readily repeatable, and she did not indicate any inconsistency in the appearances of the lights. We'll examine her sightings in detail in *Chapter 6 – "Classic" Marfa Lights* and *Chapter 7 – Lights Across the Alamito.*

---

[4] *Classic,* in this sense, means the view from the MLVC south-southwest showing the lights of vehicles on US67 headed toward Marfa.

Probably the publication that is most quoted as being the first to discuss mysterious lights at Marfa is the one in *Coronet* magazine in 1957.[9] Indeed, the article does discuss a mystery, and it clearly refers to the lights seen in the direction of the Chinati Mountains by travelers on US90 between Alpine and Marfa. Additionally, the *Coronet* article reviews some of the folklore and popular explanations for the lights. But, it doesn't provide any reports from first-hand observers. The one-page article fits squarely in the category of a *folklore* review.

In 1976, the first serious treatment of the Marfa lights appeared. The book, *Tales of the Big Bend,* was written by an English professor, Elton Miles, at Sul Ross State University in Alpine, Texas.[10] It was a serious treatment, but not one that provided any first-hand reports. Indeed, true to the title of the book, it treated "tales" of the region. Chapter 11 of Miles' book is entirely about the Marfa lights from a folklore perspective. He provided, not only a description of the range of myths about lights, but also discussed how the stories changed over time.

Miles noted that myths tend to reflect the concerns of people at the time. Thus, stories from the end of the 19th century associated the mysterious lights with Indians, whereas stories from the 1940–1950 period were more likely to be connected to German soldiers or classified U.S. research projects. The book is a delight to read, even today. It contains a photograph of the "classic" Marfa lights. This photograph, taken by a local rancher, is likely to be the first, widely distributed, published representation of the classic Marfa lights.

Judith Brueske made a major contribution to the truth about Marfa lights with her book, *The Marfa Lights,* first published in 1988.[11] She set out to go beyond folklore and find people who could attest to having personally seen mysterious lights. Additionally, she asked if they had seen them up close.

The stories she captured of sightings "up close" to the observer are unique from the mass of literature on mysterious lights. The quality of her contribution sets it apart from all the rest. For the first time, one hears of lights that were a foot or two across and sometimes varied in color. These features are in sharp contrast to the typical sightings of tiny lights that are comparable to a bright star.

Another point made by Brueske is that the mysterious light sightings are not confined to the Marfa plain. She listed a number of places in the region where people have reported seeing mysterious lights. Later (2004), in a "Letter to the Editor" of *The Big Bend Gazette,*[12] she described an even larger geographic expanse over which mysterious lights have been reported —

> *... from along the Alamito Creek that runs between Marfa and the Rio Grande, from both the Texas and Mexican sides of the Rio Grande itself (from Lajitas and downstream at least as far as Redford), from in the Chinati Mountains and the foothills east of them, from Shafter and from the flat between Shafter and Presidio, from the flat on Highway 118 south of Fort Davis, from between Fort DBMS*[5] [sic] *and Marfa, from up around Saragosa and Grandfalls, and, of course, from the official Viewing Site on Highway 90 between Marfa and Alpine that overlooks Mitchell Flat and the various mountains beyond.*

The reports given to Brueske were *first hand*. But much of the material in her book is paraphrased making those portions *second-hand reports* to the reader. Nevertheless, Brueske's

[5] The letter in *The Big Bend Gazette* had the words *Fort DBMS*, but they probably meant the words to be *Fort Davis*.

book contains the best descriptions of sightings that are truly mysterious. It is a must-read for anyone interested in the Marfa lights. We will devote three chapters to analyzing reports of mysterious lights from her book — *Chapter 10 – Lights on the One O One Flat, Chapter 11 – Aragon ~ Lights Below the Chinatis,* and *Chapter 12 – Diamond Dust ~ Bearer of Moonbeams.*

James Bunnell added his first contribution to the literature on Marfa lights in 2003 with a book, *Night Orbs*[13] He followed that effort with a DVD entitled *Marfa Lights Video* in 2005,[14] and published a second book, *Hunting Marfa Lights*, in 2009.[15] *Night Orbs* made two major contributions to the literature on mysterious lights. First, it added photographic evidence of sightings. Now readers can examine photographs and decide for themselves how mysterious the lights are. The second contribution Bunnell made was to show how exceedingly rare truly mysterious sightings are.

Bunnell was able to produce photographic evidence through major personal endurance by visiting the Marfa plain with cameras at ready and simply waiting for long hours. The authors have put in some long, cold hours at the MLVC at night and have to commend Bunnell for his tenacity in pursuit of photographs of mysterious lights.

He also put together an automatic, nighttime, photographic system and placed it on the Marfa plain (probably on the Kerr Mitchell ranch). This equipment takes time-lapse photographs all night long without any human involvement. From his extensive library of time-lapse photographs, Bunnell was only able to assemble a few representations for his book and video — illustrating once again the rarity of mysterious lights.

The photographs in Bunnell's books introduced a new dimension to the "mystery" of a light sighting. Now, we have photographs, some without narrative, that are basically white spots on a black background. Without any visible terrain for context, interpretation is left to the viewer. Many of the photographs were taken at very high magnification, and "artifacts" caused by magnification introduce sources of confusion in these long time-exposure photographs. For example, any vibration or movement of the camera will show up as movement of the light on the photograph.

The absence of contextual clues in these black-and-white photographs can lead to misinterpretation. For instance, suppose we told you that we know of places in the deserts of New Mexico, Arizona, and Nevada where we've seen lights go up and down. You are likely to think that's rather mysterious. If we then added that there are airports at these places, you'll probably change your mind because we added the context for the sightings. Now, it makes sense to you that lights would be seen going up and down at those locations. The lights are on airplanes.

The authors believe this is what is happening in the photograph on page 42 of *Night Orbs*. The photo — taken between 8:24 p.m. and 8:56 p.m. on March 29, 2003 — shows an initial light at about 114° magnetic bearing (see *Glossary*) from Bunnell's automated photography system. In the text of his book, he states that the light divided, and one of the two lights then went up in the air and curved off to the left. He further added that the "trajectory of the fly-away was rocketlike".

What the text doesn't say is that there is an airport at that location. The runway is oriented almost straight along the path from Bunnell's camera to the airport. Thus, if a plane took off away from the camera,[6] at first all that the camera would detect is the light going upward since the background is black. As the plane turned left, the spot of light would then appear to be both rising and going left. This is exactly what the photograph shows. What appears to be mysterious has a plausible explanation.

The discussion above shows how vitally important it is to have the background be visible in the photograph; the background puts the light in context. The need for context is one of the reasons we devoted so much material in the second chapter to developing the setting of the Marfa plain with its many ranches and roads. Though landing strips are numerous on the Marfa plain, they were not shown in Figure 2–1 because most of them are little more than a cleared strip of land. The landing strip mentioned in the paragraph above is considerably more than just a clear strip of land. It has its own FAA designation, 0TA7, and is located at a major ranch, owned by a company in San Antonio. On the map in Figure 2–1, the airport is at the ranch shown near the center of the figure, just above, and to the right of the last *e* in the word *Antelope*.

Even without knowing the details of the Marfa plain, people seeing the photographs in Bunnell's book would likely think that they show cars and trucks driving on ranch roads on the plain. The traces of light are usually straight and parallel to the ground, and seem to be correlated to the arrangement of roads in Figure 2–1.

Two general claims set forth by those seeing mysterious

---

[6] The winds were from the northeast the night of March 29, 2003; so a plane would have taken off flying eastward, away from the camera. It would likely have turned into the wind (which means left) to gain altitude.

lights is that they merge and divide, or go up and down. Bunnell, and others, have added that the lights tend to pulse in intensity. These observations are intriguing and warrant detailed review. We will examine these unusual light behaviors in *Chapter 6 – "Classic" Marfa Lights, Chapter 7 – Lights Across the Alamito, Chapter 10 – Lights on the One O One Flat,* and *Chapter 11 – Aragon ~ Lights Below the Chinatis.*

The most definitive work explaining one category of Marfa lights comes from the Society of Physics Students at the University of Texas at Dallas (UTD)[16] The category they examined is what we've referred to as the "classic" Marfa lights. These lights can be seen from the MLVC at locations just west of the flashing red radio tower light. (See Figure 2–1.)

The students took video recordings of the "mysterious" lights on nights between May 10 and May 14, 2004. Then, they digitized the videos and transferred them to a computer for analysis. Upon high-speed playback of the images, it was clear that the lights all followed the same path from left to right, and that they all "went out" at the same places along the path. In their paper, the students provided a link to the videos they produced. Regrettably, that link no longer works. But, we downloaded their videos in 2006 and were thoroughly convinced that the lights followed the same path every time.

The UTD students also parked one of their vehicles alongside US67, at a location where lights had been seen. They established radio communication with the video crew at the MLVC, and then blinked their headlights on and off. The video crew confirmed that they saw the blinking headlights. Through their radio link, the students established that when other vehicles passed their parked car on US67, the passing vehicle's headlights were observed simultaneously at the MLVC.

In their *first-hand report* on the experiments, the UTD students concluded absolutely that the lights they observed — the ones we term "classic" Marfa lights — were the headlights of vehicles traveling on US67 toward Marfa.

From this collection of mysterious light stories, we've selected intriguing examples of light phenomena for more detailed examination in this book. The subjects are listed in the table below.

Witness and Location	Description	Chapter
Ellison [1] Antelope Springs	Mountain moved closer to him	9
Kershner [8] MLVC	Lights seen southwest of MLVC	6
Kershner [8] Marfa, Texas	Lights seen east of town from Jordan's barn	7
Kershner [8] MLVC	Lights appeared as streaks	7
Various Observers Several Places	Lights pulsed, divided, and merged	6, 7, 10
Wagers US90	Mirage that moves sideways	8
Black [11] Goat Mountain	Yellow-orange lights above creosote bushes	10
Reed [11] Aragon, Texas	Lights that go up	11
Olsen & Sowder [11] US90	Colored balls one above other, seen in car mirror	12

Each of the examples is different, and the descriptions of the observations could be read in any order. But the technical basis for the explanations of the phenomena builds from one chapter to the next. So be prepared to learn terminology and scientific facts throughout the book that will be used in subsequent chapters.

We begin in the next chapter, *Bringing Lights to Marfa,* by looking at the historical development of light on the Marfa plain.

—§—

3—Sources

[1] Cecilia Thompson, *History of Marfa and Presidio County, Texas 1535-1946, Volume 1, 1535-1900* (Marfa, TX: The Presidio County Historical Commission, 1985), 194.

[2] Michael Hall, "The Truth is Out There," *Texas Monthly*, June 2006, 126-127, 258, 261-266.

[3] Thompson, 197.

[4] O.W. Williams. *Pioneer Surveyor–Frontier Lawyer, The Personal Narrative of O. W. Williams, 1877-1902*, ed. S.D. Myres (El Paso, TX: Texas Western Press, The University of Texas at El Paso, 1968), 222-249.

[5] Oscar Waldo Williams, "Alsate," in *By the Camp-fire in the Southwest* (Dallas, TX: Martin Stationery Co., 1934), 8-13.

O. W. Williams, *Alsate, The Last of the Chisos Apaches* ([n.p., n.d.]), 1-8.
[The version of the pamphlet the authors consulted is among the holdings of Southern Methodist University's DeGolyer Library in Dallas, TX; the call number is E99 A6 W55.]

[6] Williams, *Pioneer Surveyor-Frontier Lawyer*, 332.

[7] Judge O. W. Williams, "Alsate – The Last of the Chisos Apaches", *Frontier Times*, February 1944, Vol. 21, No. 5, 185-189.

[8] Deric Bownds. "Marfa Stories told by Ruth Bownds Kershner", January 8, 1999. (Compiled by Susan Kershner Mundine, November 2001). Available from: Deric Bownds,

http://www.dericbownds.net/Bownds_History/Marfa_Stories/MarfaStories.htm

[9] Paul Moran "The Mystery of the Texas Ghost Light", *Coronet*, July 1957, 57.

[10] Elton Miles, *Tales of the Big Bend* (College Station, TX: Texas A&M University Press, 1976), 149-167.

[11] Judith M. Brueske, *The Marfa Lights — Being a collection of first-hand accounts by people who have seen the lights close-up or in unusual circumstances, and related material,* revised ed. (Alpine, TX: Ocotillo Enterprises, 1989).

[12] Judith M. Brueske, "Letter to the Editor", *The Big Bend Gazette,* (September 2005) Vol. 5, No. 9, 12-13.

[13] James Bunnell, *Night Orbs* (Cedar Creek, TX: Lacey Publishing Co., 2003).

[14] James Bunnell, *Marfa Lights,* DVD. Video presentation by James Bunnell. (Cedar Creek, TX: Lacey Publishing Company, 2005.)

[15] James Bunnell, *Hunting Marfa Lights* (Benbrook, TX: Lacey Publishing Co., 2009).

[16] Society of Physics Students, "An Experimental Analysis of the Marfa Lights," December 10, 2005, University of Texas at Dallas, http://www.spsnational.org/wormhole/utd_sps_report.pdf [The link in the PDF to the videos they produced no longer works.]

4 — Bringing Lights to Marfa

LIFE IN THE BIG BEND REGION during the middle of the 1840s continued at much the same pace as it had for hundreds of years. Most activities took place during daylight hours when there was ample sunlight. If nighttime actions were required, people would try to plan them around the presence of moonlight. But, moonlight wasn't a light source early inhabitants could control or take into a shelter. For that, they made fire.

In fact, fire was very likely the only manmade source of light created by the inhabitants of the Big Bend at this time. Fire was a vital source of nocturnal light. But more than that, fire provided warmth, a means for cooking food and brewing intoxicants, and a signaling ability for communication between locations. Fire was so precious that early travelers and migrant

Indians were known to carry smoldering roots or embers from a previous fire and blow on them to ignite fires when they stopped for the day. Settlers made efforts to ensure that their hearth fires never went out completely. Without fire, life was very difficult indeed.

But the region was about to be changed forever. During the next 40 years, nearly all types of light would be introduced. By the time of the legendary birth of the Marfa lights in 1883, there would be a rich collection of sources to produce the mysterious lights. In the next few pages, we'll draw a thumbnail sketch of the forces that opened up the Big Bend and brought about the introduction of varied light sources.

By the mid 1840s, none of the "modern" sources of light had arrived in that desolate no-man's land east of the Rio Grande and west of the Rio Pecos. We say *no-man's land* because the region was in dispute. The Republic of Texas claimed all territory westward to the Rio Grande, whereas, Mexico asserted that the boundary was the Nueces River, which is even farther east than the Pecos. In fact, neither the Texans nor the Mexicans had settled the region between the Rio Grande and the Rio Pecos. It was essentially all occupied by Indians.

Indians had inhabited the region for centuries. Jumanos had communities along the Rio Grande near the current city of Presidio; Apaches occupied the mountainous regions on the northeastern side of the Rio Grande; and Comanches traveled freely from the plains of north central Texas though the Big Bend on their annual campaigns to plunder the towns and rancherias of Mexico.

Over the centuries, Indians traversing the Big Bend developed well-beaten trails. Figure 4–1 shows the major trails and springs in the Big Bend in the 1840s.

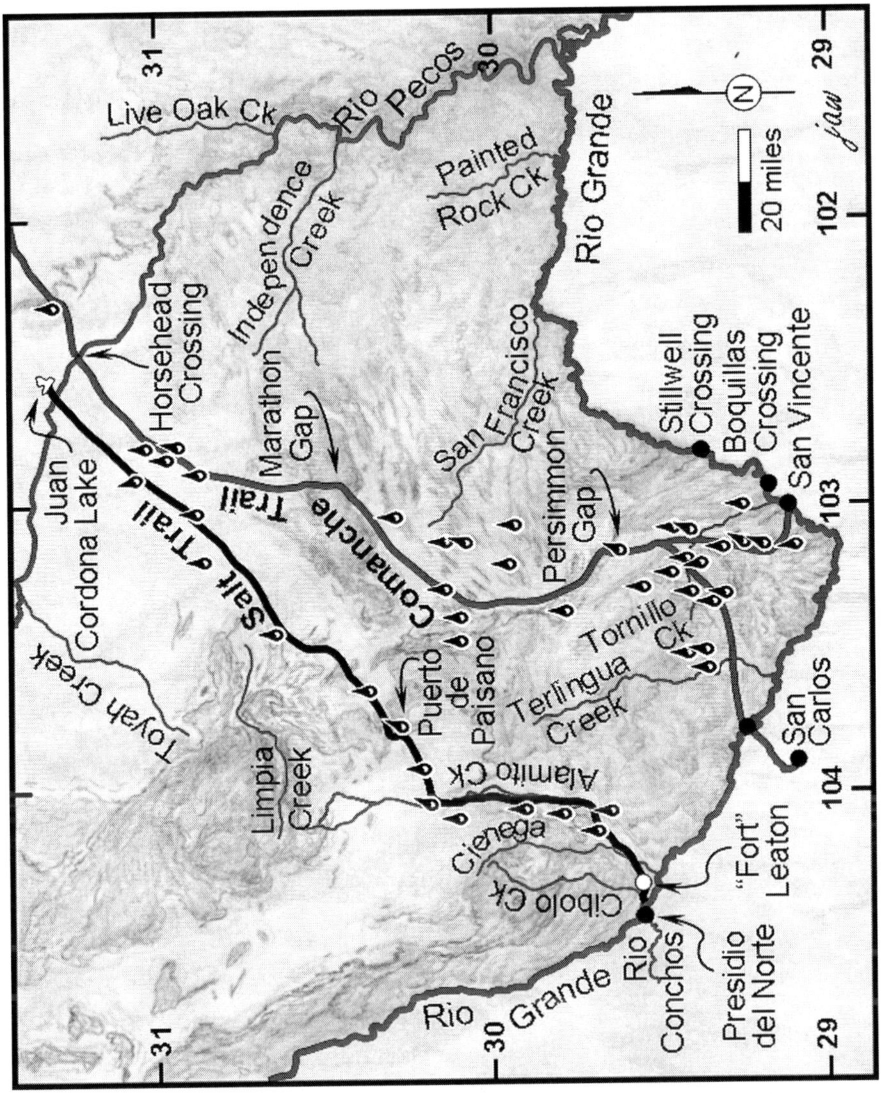

Figure 4–1. Major trails and water sources (shown as raindrops) in existence in the Big Bend of Texas at the middle of the 1840s.

The Salt Trail was used for commerce; the Comanche Trail for warfare. Along those trails, water was available to those who knew where and *when* to find it. That was no accident.

For, without water, there wouldn't be a trail! Availability of water and passages through the mountains determined where people traveled in the Big Bend.

The oldest of the trails shown in Figure 4–1 was the Salt Trail. It extended 190 miles northeast from the settlement at the confluence of the Rio Grande and the Rio Conchos (which flows north from Mexico). Settlements have existed at the Rio Grande and Rio Conchos for more than 10,000 years. And, for centuries Native Americans have walked the length of the Salt Trail to hunt buffalo northeast of the Pecos River crossing, and to collect salt from the salt lake now called *Juan Cordona Lake*, in Crane County, Texas. After killing their buffalo, they would salt the meat at the lake and then haul their harvest back down the trail.

The Spanish explorer Antonio de Espejo is reported to have encountered three Jumanos Indians at the salt lake in 1583.[1] Espejo accompanied these Indians on their 12-day journey back down the Salt Trail to their settlement on the Rio Conchos.

The first U.S. citizen who is documented to have taken a wagon train over the Salt Trail was Henry Connelly, a trader in Chihuahua, Mexico. In 1839, he traveled from Chihuahua to the Presidio area where he picked up the Salt Trail and followed it north until he veered off to cross the Rio Pecos at Horsehead Crossing. He then pressed onward to Arkansas to purchase supplies for resale in Chihuahua. The following year, he retraced his path to Chihuahua. Because of his trip over this trail, early maps labeled it the *Connelly Trail.*

By 1852, the U.S. Army had surveyed the Salt Trail and made improvements to the road to facilitate wagon traffic.[2]

This was fortuitous for a correspondent for the *New York Times*, who was located in Chihuahua at the time. When it came time for him to return to New York, Julius Froebel decided to accompany a wagon train carrying Mexican pesos to the United States.

His train of seven wagons and 100 mules left Chihuahua in May of 1853 proceeding north to Presidio del Norte. Crossing the Rio Grande, they continued along the old Salt Trail until it intersected the newly-created San Antonio to El Paso Road. At the junction, they turned eastward to San Antonio, and then down to the Gulf coast at Indianola. He recorded his trip and locations in a diary, and later published the material in a book covering his seven years of travel.[3] All of the watering sites and major features along the Salt Trail are chronicled in his book. The trail and the features along the way were all well known to the leaders of the wagon train. In fact, the watering sites already had well-known names that are essentially the same ones used today.

The Salt Trail was to be pressed into major commercial use by the early freighters that operated between San Antonio and Chihuahua during 1850–1883, especially following the end of the Civil War (1865). The Chihuahua silver mines and surrounding communities needed goods manufactured in the eastern United States and Europe. Being in possession of some of the richest mines of the time, they had silver to spend. As a consequence, a trail came into existence from Indianola, on the Gulf Coast of Texas, up to San Antonio, northwest for 300 miles over to the Salt Trail, down to Presidio, and south to Chihuahua. This became known as the *Chihuahua Trail.*

The other great historical trail shown on Figure 4–1 is that attributed to the Comanche Indians. The Comanches would set out on this trail when the moon became full in the time of year around September. They traveled into Mexico where they captured slaves, cattle, horses, and any other valuables of their choice. Women and children were enslaved; men were killed. Foot traffic of slaves, horses, and cattle on these trails scarred the earth so indelibly that the trails were easy to detect and follow.

Not all Mexican settlements were treated alike, however. The Comanches would make treaties with some towns or states in Mexico, but wage war on others. Those communities protected by treaties traded with the Indians for goods. Two sanctuaries just across the Rio Grande were San Vicente and San Carlos (just east of the current town of Manuel Benavides).

The Comanche Trail shown in Figure 4–1, with its split at the bottom, is just one example of the paths the Indians followed. In fact, they followed many variations through the Big Bend region. But all the trails tended to go through Comanche Springs (current Fort Stockton) and the region around Marathon with its rich collection of springs. From there, the trails took their separate paths to reach the crossings on the Rio Grande at Lajitas (San Carlos crossing) or the San Vicente area.

These historical trails in Figure 4–1 illustrate how primitive the Big Bend was in the late 1840s. But things were about to change. President John Tyler signed legislation setting 29 December 1845 as the date of annexation of the Republic of Texas into the Union. That set in motion a great instability between Mexico and the United States. The dispute over the western boundary of Texas was to be resolved by armed conflict. But the timing for the conflict was ill-considered by the Mexicans. Now, instead of fighting a war with the Republic of Texas, Mexico would be at war with the entire United States.

The Mexican–American War lasted from 1846 to 1848, and ended with the United States occupying not only the disputed region west to the Rio Grande, but the entire western half of the present contiguous 48 United States. Even the capital city of Mexico was captured. The Mexican government was forced to cede Upper California, most of present-day Arizona and New Mexico, and the disputed land between the Nueces and the Rio Grande — including the Big Bend region. This settlement was agreed to by the heads of the Mexican and United States governments in the Treaty of Guadalupe Hidalgo.[1][4]

Article V of the Treaty of Guadalupe Hidalgo called on the governments of Mexico and the United States to each appoint a commissioner and a surveyor to "run and mark the said boundary in its whole course (from the Pacific Ocean) to the mouth of the Rio Bravo del Norte" (Rio Grande). While the negotiations of the commissioners and the efforts of the surveyors started on schedule, difficulties with the treaty immediately arose. It seems that the treaty referenced specific landmarks (like El Paso) as well as an 1847 map of the United States of Mexico by J. Disturnell.[5] The map gave incorrect locations for relevant landmarks. Those errors led immediately to the question: Do the surveyors accept the landmarks' true locations on the earth as defining the border? Or, do they accept the border as defined by Disturnell's map coordinates?

Years of negotiations ensued, culminating with the ratification of the Gadsden Purchase Treaty on December 30, 1853.[6] In this treaty the boundaries between Mexico and the states of the United States were moved south to their current locations.

---

1 Guadalupe Hidalgo was a town outside of Mexico City where the treaty was negotiated on 2 February 1848. The treaty was proclaimed on 4 July 1848.

This readjustment greatly increased the size of the regions that would become present-day Arizona and New Mexico. It also provided a passage for a train track around the mountains north of present-day El Paso.

While all the treaty/boundary negotiations were proceeding, the area was opening up to settlement and exploration. Fort Leaton (shown in Figure 4–1) was among the first trading posts established in the Presidio area following the Mexican–American War. It began as a Spanish fort in 1773, and went through several owners before Ben Leaton bought the property in 1848 and built on the ruins of the Spanish fort. The name, *Fort Leaton*, suggests that it was part of the U.S. Army fort system; but that wasn't the case. Fort Leaton was actually the home, trading post, and personal fort of Ben Leaton. Though Ben Leaton died in 1851, and his wife gave up the property shortly after, Fort Leaton continued to appear on maps until the late 1890s. Today, the site is Fort Leaton State Historic Site.

Subsequent to the Mexican–American War, the United States was also funding comprehensive reconnaissance activities in the territories they were about to receive. These efforts included not only exploration by the U.S. Army, but surveying and mapping by the U.S. Army Corps of Topographical Engineers. The immediate interest of the United States was a road from San Antonio through the Big Bend region to El Paso, and beyond to California. The longer term objective was to have a railroad across the southern states that would link the east and west coasts. Such a railroad would enjoy year-round operation by avoiding the snows further north.

Army Lieutenant William Whiting and 15 others were given orders to seek a direct path between San Antonio and El

Paso. They left San Antonio in February 1849 and did not return for 104 days — long after they had been given up for lost or dead. They became the first of the government explorers to pass through the Davis Mountains. Their trail led them through Wild Rose Pass and to a small stream in the mountains (Limpia Creek) at a place that would later become Fort Davis. They recommended that the Army build a road through the region and on westward. Figure 4–2 shows where Fort Davis was ultimately built.

Today, the Fort Davis National Historic Site has markers in place showing the route the road followed as it passed by the fort. The basic trail discovered by these hardy explorers would go on to become the San Antonio–El Paso Road. Intersecting the Salt Trail near Comanche Springs (later the site of Fort Stockton), this road was the standard for southern transportation across west Texas to El Paso.

By the early 1850s, the road was a major corridor for emigrants headed for the California gold fields. But travel and development were still hampered by Indian raiders throughout the Trans–Pecos region.

By 1854, it was clear that the Army needed a series of posts in this region to protect the route westward. In 1854, Fort Davis was established, followed by Fort Lancaster the following year. Fort Quitman (off the left edge of Figure 4–2) was added in 1858; Fort Stockton at Comanche Springs was the last of this chain of forts to be added in 1859.

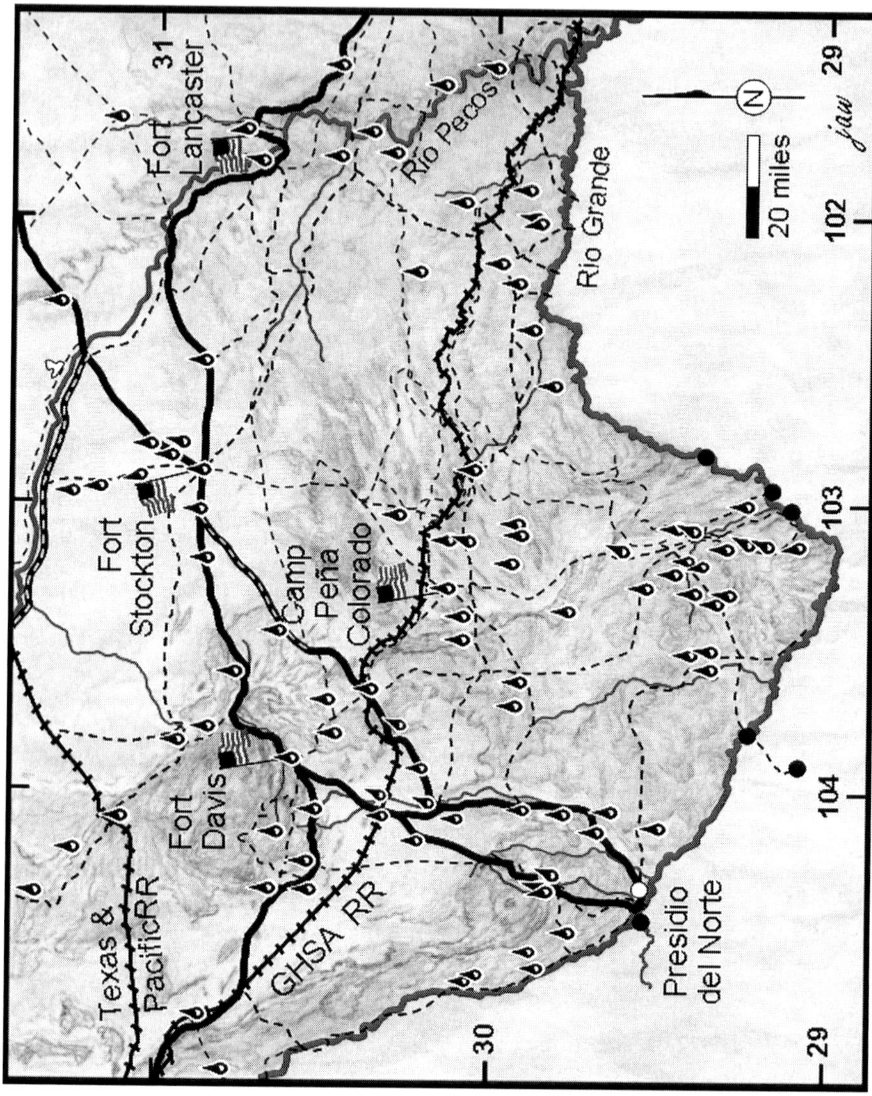

Figure 4–2. Complex of roads, trails, and railroads in the Big Bend region in 1883. Black roads show trails that grew into arteries of commerce (including immigration). Dashed lines show trails that were carved out through government exploration. The railroads may have followed existing trails in some places but they were planned to achieve the larger goal of transcontinental connectivity. The mountains that would later be named the Davis Mountains, lie under the label *Fort Davis* in the figure.

With the completion of these forts, the San Antonio–El Paso route passed through Fort Lancaster, Fort Stockton, Fort Davis, and out onto the valley west of the Davis Mountains. From there, it angled north, skirting alongside the Davis Mountains before crossing the valley to the last watering location shown on the map — Van Horn's Wells.[2]

Off the map in Figure 4–2, the road continued on westward. It worked its way past the Eagle Mountains, over Quitman pass (just south of Sierra Blanca on Interstate 10), and down to Fort Quitman on the Rio Grande. From there, the road followed the Rio Grande northward on its final leg into El Paso, along what is now RM192.

A comparison of Figure 4–1 to Figure 4–2 shows that Fort Davis was not on the old Salt Trail. To satisfy the need for agricultural supplies to the fort from growers along the Rio Grande near Presidio del Norte, freighters split off from the old Salt Trail at Tinaja de San Esteban (see Figure 2–1) and developed a trail straight up to the fort. Later, when the Galveston, Harrisburg and San Antonio Railroad passed through the region, it would cross the road to Fort Davis at the location that would become Marfa.

Though Indians were a constant threat in the 1850s, the forces of the U.S. Army were too much for the Indians to resist; and by about 1880, the Indian threat was brought under control. The U.S. Army did not conduct further significant operations against Indians of the Trans–Pecos.

Travel westward, particularly to the gold fields of Califor-

---

[2] There is a historic marker for Van Horn's Wells on US90 about 10.7 miles south of Van Horn, Texas. Just 3.1 miles north of the marker, the Southern Pacific Railroad crosses over US90.

nia, provided a major force in opening up the Big Bend region. The railroad, which was to come, would drive economic development even more. The Southern Pacific Railroad Company initiated their development of an all-weather route connecting the oceans on the two sides of the United States in 1877. They advanced track from California into Yuma, Arizona and continued eastward. In 1881, they reached El Paso, Texas, where they were obliged to stop laying Southern Pacific tracks. It seems that in that time no one could operate a railroad company in Texas unless the company was chartered in Texas.

But the Southern Pacific hadn't blundered into a problem with Texas. They had already arranged for a partnership with the Galveston, Harrisburg, and San Antonio (GHSA) Railroad. GHSA had a Texas Charter, and the crews that had laid Southern Pacific tracks eastward to the Texas border, simply continued on into Texas laying GHSA tracks. As the tracks out of El Paso advanced eastward, other tracks of GHSA were extended from San Antonio westward to meet them.

The two sets of tracks met west of the mouth of the Rio Pecos. A silver spike was driven on January 12, 1883, officially joining the two railroads.[7] The first passenger train was scheduled to leave New Orleans on Tuesday, January 30, 1883,[8] passing through the Big Bend area, bound for San Francisco — 2,127 miles away. High-speed travel[3] of passengers and cargo between the Gulf of Mexico and the Pacific Ocean was now possible year-round.

---

[3] The concept of "high speed" has to be measured relative to the previous transportation options available to travelers. The train was certainly faster than an ox cart, which might cover only 20 miles a day. In 1886, just four years after the GHSA Railway crossed Texas, the 632-mile trip from San Antonio to El Paso would take a little over 32 hours,[9] for an average speed of about 20 miles per hour.

The GHSA railroad would vastly accelerate the development of the Big Bend. Tracks reached the Marfa–Murphysville (now Alpine) region in early 1882,[10] and both places became watering stops on the route. Other stops in the region included: Valentine, Ryan, Aragon, Nopal, Paisano, Strobel, and Marathon. The watering stations used by the steam locomotives are gone now. However, if you look on the north side of the road when traveling on US90 between Valentine and Paisano Pass, you can still see metal sheds along the tracks displaying the names of the old stops.

By 1883, all the government-sponsored exploration,[4] immigration into the area, and exploitation of the trails for commercial purposes had opened up the Big Bend region. As shown in Figure 4–2, what had been a desolate unexamined land with a few trails, now was a network of trails, roads, and two railroad lines.

All this travel into, and through, the Big Bend brought lights! They came in people's pockets, on stage coaches and trains, and in freight wagons. People needed light for nighttime activities and where people went, lights followed. Lights were used in places we wouldn't expect in that time.

One of the freighters operating between San Antonio and Chihuahua, August Santleben, tells a story of returning from Chihuahua in 1874.[12] On this trip, he had as passengers "Pro-

---

[4] The Army Corp of Topographical Engineers created extensive maps documenting the military's explorations. In addition to the expeditions led by Colonel Whiting and Major William H. Emory,[2] two of the more famous were the camel expeditions led by Brevet Second Lieutenant William H. Echols and Second Lieutenant Edward L. Hartz prior to the Civil War. Their diaries documented these trips and were later published as reports to Congress.[11]

fessor Manuel Manso and his orchestra group". Santleben recounts having dances at night on two occasions — once just north of Presidio del Norte and once outside of Fort Stockton.

> *We always selected a camping place with a level space suitable for dancing, and every night the canvas was spread inside the corral, which was illuminated by the light of* ***candles*** *placed on the wagon wheels, and we danced, to the sweet music discoursed by Manso's fine orchestra, to the limit of endurance.*[12]

We remarked at the opening of this chapter that fires were probably the only manmade lights in the Big Bend prior to the middle 1840s. All other lights — except astronomical ones — were transported to the area. By 1883, the list of manmade light sources had grown from just campfires to include matches, candles, and oil-fired lanterns. Table 4–1 lists many light sources capable of producing illumination after dark. The table also shows when these sources were likely to have been introduced to the Marfa plain.[5]

None of these light sources seems mysterious. In fact, they all seem quite ordinary and common. Consider the moon, for instance. Surely no one would think that the moon's light is mysterious. Or, would they?

Well, the truth is that under the correct circumstances the moon can produce a very mysterious appearance. And, if the observers' orientation relative to the moon is right, they can mistake the moon for something very different from the lunar

[5] Obviously, astronomical sources of light, such as the moon, have provided light on the Marfa plain for as long as there has been a Marfa plain. But in the earliest days, there were no human inhabitants to witness these lights or to report a mysteriously appearing light.

orb. Such a situation is described in *Chapter 12 – Diamond Dust ~ Bearer of Moonbeams,* where the moonlight is contrasted to witness reports of mysterious lights seen from US90 one night in May of 1987.

LIGHT SOURCE	**ARRIVAL ON MARFA PLAIN**
Moon	Eons ago
Planet Venus (called a *star* by some)	Eons ago
Polaris (North Star)	Eons ago
Campfires	Centuries ago
Matches	About 1850
Candles	About 1850
Oil lamps (for home use)	Mid 1800s
Train kerosene headlamps	1882
Train carbon arc headlamps	About 1908
Automobile kerosene headlamps	About 1908
Automotive incandescent headlamp	1912
Electric lights in buildings	1907
Mercury vapor lamps (MVL)	Mid 1900s

Table 4–1. Light sources available on the Marfa plain over the course of time.

If the moon can be perceived as mysterious under the right set of conditions, then it is likely that other light sources can produce mysterious effects as well. But for manmade sources to be categorized as *mysterious,* they need to be far enough away that the light source itself isn't identifiable.

At what distance is a manmade light still visible? As calculations in the next chapter — *Illuminating the Mystery* — will demonstrate, the distances can be surprisingly large.

For example, a surveyor's electric light, prepared for the United States Coast and Geological Survey (USCGS) in the mid-1920s, was seen at a distance of 152.9 miles with the unaided eye! Prior to the electric light, the USCGS used acetylene lamps. Two of these automobile acetylene headlights were observed at a distance of 133.9 miles at Pilot Peak, Nevada.[13]

To put this in perspective, imagine you are at Cape Canaveral, Florida, standing on a high tower right beside the Atlantic Ocean, and looking straight west across the entire peninsula of Florida. The Gulf of Mexico is about the same distance away from you as the observation distance at Pilot Peak, Nevada.

Lights of more modest power, such as matches, candles, campfires, bonfires, and oil-fired lanterns were present in west Texas — and on the Marfa plain — long before 1883 when Robert Ellison unloaded his cattle in Murphysville (now Alpine) and herded them through the Paisano Pass.[14] Gunpowder flashes were also seen on occasion. In fact, surveyors mapping the United States–Mexico boundary (1849–1857) for the Treaty of Guadalupe Hidalgo and the Gadsden Purchase[15] used signal fires and gunpowder flashes in the Big Bend area to establish some of the survey lines.[16]

The common match made its appearance in the Big Bend around 1850.[6] We know this from the writings of early explorers of the region. For example, Jose Policarpo Rodriguez trav-

---

[6] By 1850, match makers in London produced 10 million boxes per year!

eled the desert southwestern region between San Antonio and New Mexico on assignments for the U.S. Army. In his memoirs, he tells of crossing the Frio River (west of San Antonio) on horseback and becoming soaking wet in the winter of 1852.

> *The weather was cold and I nearly froze. I was wet above the waist, but my matches were dry. I rode up to a large Spanish dagger (Yucca), which was covered with dry, dead bark and leaves. I struck a match to it and it burned like tinder. I warmed by it, as did my horse also.*[17]

It seems likely that the Indians in the Davis Mountains also made use of matches from a story found in the diary of Lieutenant Whiting in 1849,[18] and the memoirs of his scout, Jose Policarpo Rodriguez.[19] Both of them tell of encountering a group of Apaches consisting of an older man, three women, and a boy west of the Rio Pecos near the Davis Mountains. During their discussions with the man, one of the Indians set fire to the grass as a call for help from their tribesmen in the mountains. Though a skilled person could start a fire in a matter of minutes using ancient techniques,[7] it seems more likely that the Indians carried matches.

Trains with oil-fired headlights were common in the late 1800s. And, when the GHSA train tracks arrived in Marfa on January 16th 1882,[20] they brought a new source of light to the northern side of the Marfa plain. These oil-fired train lights illuminated the Marfa plain from time to time until the late1880s when carbon arc lamps were introduced. The carbon arc lights were much more powerful and projected light beams that could be seen from many miles away.

---

[7] Fire was produced using tinder with steel and flint, or by twirling a spindle of wood against a wooden base.

Later, trains brought light out onto the Marfa plain following completion of the extension of the Kansas City, Mexico, and Orient (KCMO) Railroad across the Marfa plain and down to Presidio in 1930. (See Figure 2–1.)

While trains illuminated places along their tracks, their commerce led to the introduction of lights far beyond the railroads. From places like Marfa and Alpine along the tracks of the GHSA, various means were used to transport rail goods to locations like Fort Davis and Fort Stockton to the north and Lajitas, Terlingua, Shafter, Presidio, Ruidoso, and Candelaria to the south.

Pack mule trains were an early source of transportation. But, at least by 1918, steam-powered tractors were in service pulling wagon trains over roads in the Big Bend region. In W.D. Smithers' book, he shows a photograph he took of a steam-powered tractor pulling nine covered wagons on a road in Pinto Canyon (between the Chinati and Viejo Mountains).[21] The road, now RM2810 or the Pinto Canyon Road, was the supply route from Marfa to Ruidoso and Candelaria. The tractor could operate at night as is evident from its headlights shown in the photograph.

By 1908, electric headlights were installed on automobiles as *optional* equipment. By 1911, they were *standard* equipment on some U.S. passenger cars.[22] Automobiles with headlights appeared in the Marfa region at least by 1910; and autos were used for passenger service between Marfa and Shafter by 1912.[23] Cecilia Thompson shows a 1914 picture of a collection of about nine autos parked side-by-side in Marfa for a meeting of the city's boosters.[24] So it's not surprising that by the end of the decade, Smithers would show a picture of a car with two large headlights being ferried across the Rio Grande all the way down at Lajitas (ca. 1916–1920).[25] As we'll see in the next chapter, these headlights were capable of projecting strong illumination.

With the exception of the moon, Venus, and stars, all the light sources discussed so far had the ability to move from place to place on the earth. But as the 19th century came to a close, another type of lighting began to emerge. These lights would be fixed in place instead of being mobile.

In 1880, Thomas Edison invented a light bulb filament which could last over 1,200 hours. The following year (1881), New York City was wired for street lights. In 1883, Galveston was the first city in Texas to wire its streets for electric lights. But, it would take some time for electricity to reach the Big Bend area. In fact, there would be communities near the Rio Grande that wouldn't have electricity until the middle of the 20th century[26] — following the enactment of the Rural Electrification Act of 1936.

Marfa, being the Presidio County seat, led the way in electrification. Electrical service was established in the town on March 16, 1907.[27] Photographs of Marfa taken around 1916 show electric power lines in place. And, the U.S. Army had electricity at its Camp Marfa facility at least by 1919.

The final entry in Table 4-1 is a relatively modern source of light — the mercury vapor lamp (MVL). Today, MVLs are commonly used on ranches and are prominently visible at night on the Marfa plain. They can be seen from enormous distances as we'll demonstrate later.

With so many lights to choose from, it is small wonder that tales of mysterious lights emerged. But it's one thing to have a light source; it's quite another to identify a light as being *mysterious.* As we said earlier, for a light to have a mysterious appearance, its source cannot be discernible. So how close do you have to be to identify a common source of light? Or put another way, how far away do you have to be before you can't tell what you're seeing? We'll examine these questions in *Chapter 5 – Illuminating the Mystery.*

—§—

4— Sources

[1] Carlysle Raht, *The Romance of Davis Mountains and Big Bend Country,* Texana Edition (Odessa, TX: The Rahtbooks Company, 1963), 33.

[2] Major William H. Emory, *Report on the United States and Mexico Boundary Survey, Vol. I.* 34th Congress, 1st Session, Ex. Doc. No. 135, Washington, D.C., 1857, p. 88.

Major William H. Emory, Boundary between the United States and Mexico, Map Number 2. Library of Congress Geography and Map Division, Washington, DC, 20540-4650 USA, 1855. Call number: G3701.F2 1855 .E6 TIL.

[3] Julius Froebel, *Seven Years' Travel in Central America, Northern Mexico, and the Far West of the United States* (London: Richard Bentley, 1859), 404-428.

[4] William MacDonald, ed., *Select Documents Illustrative of the History of the United States 1776-1861* (New York: MacMillan, 1898), 365-372.

[5] J. Disturnell, *Mapa de los Estados Unidos de Mejico,* Revised Edition, (New York, 1847). [Available from the Library of Congress.]

[6] William MacDonald, ed., 390-395.

[7] "Connected. The Silver Spike and the Southern Pacific and Sunset Railroads," San Antonio *The Evening Light,* January 12, 1883, p. 1, col. 2.

[8] "Through to California", *The Galveston Daily News,* January 24, 1883, p. 7, col. 3.

[9] "Southern Pacific Company", in *The Rand-McNally Official Railway Guide, and Handbook, for the United States and the Dominion of Canada,* No. 3, September 1886 (Chicago: National Railway Publication Company, 1886), 294.

[10] "El Paso. Notes on Railroad Building—The Southern Pacific—Mexican-Central—Ice Factory, Special Telegram to The News," *The Galveston Daily News,* January 22, 1882, p. 1, col. 8.

[11] William H. Echols, "Diary", Report of Secretary of War, in Message from the President of the United States to the Two Houses of Congress of the 36th Congress, 2nd Session, Senate Executive Document No. 1, Vol. II., Washington, D.C., 1860, p. 36-50.

Edward L. Hartz, "Diary", Report of Secretary of War, in Message from the President of the United States to the Two Houses of Congress of the 36th Congress, Executive Documents of the United States, 1st Session, Vol. II, No. 2, Part 2, Washington, D.C., 1859, pp. 423-441.

[12] August Santleben, *A Texas Pioneer,* (New York: The Neale Publishing Company, 1910), 171.

[13] NOAA, "NOAA Celebrates 200 Years of Science, Service, and Stewardship," National Oceanic and Atmospheric Administration. Available from: http://celebrating200years.noaa.gov/magazine/signals/welcome.html

[14] Cecilia Thompson, *History of Marfa and Presidio County, Texas 1535-1946, Volume 1, 1535-1900* (Marfa, TX: The Presidio County Historical Commission, 1985), 193.

[15] Emory, *Report.*

[16] Paula Rebert, *La Gran Línea* (Austin, TX: University of Texas Press, 2001), 32.

[17] Jose Policarpo Rodriguez, *The Old Guide* (Dallas, TX: Publishing House of the Methodist Episcopal Church, [n.d.]), 54. (Dictated to Rev. D.W. Carter, D.D., from 1892 to 1897.)

[18] W.H.C. Whiting, "Whiting Diary", *Publications of the Southern History Association* 10, no. 2 (March, 1906): 78-96.

[19] Rodriguez, 23-25.

[20] S.G. Reed, *A History of the Texas Railroads* (Houston, TX: St. Clair Publishing Co., 1941), 197.

[21] W.D. Smithers, *Chronicles of the Big Bend* (Austin, TX: Texas State Historical Association, 1999), 28.

[22] David W. Moore, "Headlight History and Harmonization," *Report No. UMTRI-98-21* (Ann Arbor, MI: The University of Michigan Transportation Research Institute, June, 1998).

[23] Cecilia Thompson, *History of Marfa and Presidio County, Texas 1535-1946, Volume 2, 1901-1946* (Marfa, TX: The Presidio County Historical Commission, 1985), 92.

[24] Thompson, *Volume 2,* 75.

[25] Smithers, 114.

[26] Cindy Edwards, "Remembering the Early Days," *Update — Rio Grande Cooperative, Inc.*, May 2003, p. 5.

[27] Thompson, *Volume 2,* 34.

5—Illuminating the Mystery

MYSTERIOUS LIGHTS have one thing in common with ordinary lights. They both require a ***source*** of light. This statement seems so obvious that it doesn't need to be made. It is undeniable that if someone sees a light at night on the Marfa plain, then something out there emitted photons; and those photons traveled from the source to the person's eyes.

But stating the obvious, as we just did, helps to focus our attention when trying to explain the mysteries surrounding the Marfa lights. Before talking about mysteries, we must deal with the fact that — **There *MUST* be a source of light.**

In much of the previous literature, the discussions focused

on the word *mysterious* and not on the word *light*. But, it is not necessary for the ***sources*** of light to be mysterious to have the light itself be perceived as mysterious to the viewer. Lights from common sources could appear to be mysterious if something occurred after the light was emitted that caused the viewer's perception of the light to be other than what was expected.

The mysterious appearance of lights, and the opinion that common light sources are not "strong" enough to be visible at great distances, have led to their dismissal without further examination.

But can common sources of light be seen at great distances? If so, what lights are visible? And, how far away are they? In this chapter, we'll examine the sources of light, their properties, and the distances at which they can be seen.

As we discussed in *Chapter 4 – Bringing Lights to Marfa*, naturally occurring lights have been available on the Marfa plain since the beginning of time. Manmade sources of light have been around for thousands of years. And, more technologically advanced forms of light production have existed for more than a century. Some of the latter lights were present on the Marfa plain at least as far back as 1883 when Robert Ellison drove his herd of cattle through Paisano Pass and out on to the Marfa plain.[1]

Throughout the 20th century, there were a variety of high-intensity light sources present on the Marfa plain. But at the time of Robert Ellison's alleged observation, the only sources available were astronomical ones, matches, candles, oil lamps, fires, and train lights.

Other than train lights, could any of these lights be bright enough to be seen from a sufficient distance that a person could ***not*** identify what produced the light?

We can't answer this question without understanding a bit about the physics of light propagation. For those who — like the non-scientist co-author — find physics foreign, we promise you won't have to become a physicist to comprehend the relevance of this discussion. And, the conclusions will amaze you.

Let's begin by defining some fundamental terms. The three words below are probably not familiar to many readers but we'll need to use them in the following discussions, and it's easy to develop a casual understanding of their meanings.[1]

Lumens —Lumens are a quantitative measure of how much light is emitted by a source without regard for where the light is projected. It doesn't matter whether light is emitted by the sun, or a match, or a bonfire, or a light bulb; the measure of light output is lumens. Look at a light bulb package today, and you'll find light output recorded in lumens.

Intensity —When lumens from a source of light are projected into a solid angle, the intensity is specified as the quantity of lumens divided by the size of the solid angle. Picture a megaphone. Now, consider shining a flashlight into the small end of the megaphone. We idealize the flashlight and megaphone system so that the light uniformly spreads out as it flows through the megaphone filling the volume with light. If the megaphone volume is big, the intensity is low because the light is spread over

1 For readers who want more information about these terms, the *Appendix* provides an extended discussion.

a large volume. Conversely, if the volume is small, the intensity is high. You can see this for yourself if you use a flashlight with an adjustable focus. Focus it tightly, and the spot of light is bright; focus it to a broader spot, and the light is dimmer. Intensity is measured in units called *candelas*.

Illuminance —Illuminance is found by considering the quantity of light (lumens) falling on a small surface perpendicular to a light ray, and dividing the lumens by the area of the illuminated surface. At first reading, this may sound like a restatement of intensity. But it's a little different. For intensity, we divided the total lumens by the size of the angle the light occupied. For illuminance, we divide the total lumens by the *area* of the surface the light is shining on. Illuminance is measured in units called *lux*.

In *Chapter 4 – Bringing Lights to Marfa*, we discussed three astronomical light sources. Their illuminances (lumens per square meter, or lux) falling on the earth's surface are shown in the second column of the upper portion of Table 5–1.

The illuminance values reveal some facts that most people would have guessed based on their personal experiences. The full moon (0.29 lux) is relatively bright; and Polaris (being 434 light years away from earth) is very dim. In fact, Polaris produces only about one millionth of the light of a full moon. Venus,[2] which is brightest just before sunrise or just after sunset, has an illuminance in between that of Polaris and the full moon.

---

[2] Even though Venus is sometimes called the *Morning Star* or the *Evening Star*, it is a planet, not a star.

LIGHT SOURCE	**ILLUMINANCE** (measured in lux)
Moon (full)	2.9E–01 (0.29)
Venus (at its brightest)	1.6E–04 (1.6 ten thousandths)
Polaris (North Star)	4.0E–07 (4 ten thousandths)

HUMAN VISION	**ILLUMINANCE THRESHOLD** (measured in lux)
At the end of civil twilight[3]	2.0E–07 (2 ten millionths)
Looking toward the horizon at night with full moon	4.0E–08 (40 billionths)
Looking toward the horizon at night with no moon	2.0E–08 (20 billionths)
At the ultimate limit	2.0E–09 (2 billionths)

Table 5–1. *At top:* Light levels on the surface of the earth from three astronomical sources.[4] *At bottom:* Four thresholds of human light perception. The illuminance values given are measured at the eye of the observer. Illuminance units are given in scientific notation and their equivalent common terminology.[5]

The eye's ability to detect a specific light is affected by any competing lights that are present. The darker the background, the dimmer a source of light can be and still be perceived. The

---

3 The end of civil twilight is taken to be about 20 minutes after sunset in Marfa.

4 Light levels for the moon, Venus, and Polaris were taken from Reference [2]. Background light levels from Reference [3] were used with Figure 5.7 of Reference [4] to obtain the perception thresholds presented here.

5 See *scientific notation* in the *Glossary* for an explanation of number format.

bottom portion of Table 5–1 contains information about the human eye's ability to perceive light under four different ambient light conditions. The illuminance levels shown in the second column correspond to the thresholds of perception, for a person with healthy visual acuity (corrected if necessary). Sources of light that are dimmer than the values shown in the table (and thus produce lower values of illuminance at the observer's eye) cannot be reliably detected by an individual.

The first condition stated in the lower section of Table 5–1 applies to the perception of light at what is generally known as *civil twilight*. There is a precise definition (see *Glossary*) for the period at dusk known as *twilight* that has to do with the angle of the sun following sunset. However, for our purposes, we'll take evening civil twilight at Marfa to be approximately the first 20 minutes following sunset. The illumination level quoted for civil twilight refers to the end of evening twilight — 20 minutes after sunset.

The level of background light varies considerably during the evening civil twilight. It starts out quite bright right at sunset, and ends at a darkness that is too dim to allow someone to read outdoors. The planet Venus becomes visible during the evening twilight. At some time during evening twilight, the background light reaches a level where the dark-adapted human eye can just see a light with an illuminance as dim as about 2.0 E–7 lux. That's 2.0 ten millionths of one lux. Weaker lights cannot be perceived with the unaided human eye.

As the sky darkens further and the only background light is that provided by the stars and moon, a person can see even dimmer lights. Looking toward the horizon on a clear night with full moonlight, a person can perceive lights with illuminances of about 4.0 E–8 lux or brighter (40 billionths of a lux).

With no moonlight, the threshold of human perception permits seeing fainter lights with illuminances of about 2.0 E–8 lux (20 billionths of a lux). Finally, under laboratory conditions where all background light is absent, the ultimate human threshold of visual perceptibility is about 2.0 E–9 lux (2 billionths of a lux). The ultimate threshold limit is a good approximation for what you'd expect on a night with no moon, heavy cloud cover, and no manmade lights.

Now, let's consider how the astronomical light levels compare to human perceptibility. Table 5–1 shows that at civil twilight the full moon emits a light level (0.29 lux) that is more than a million times greater than our ability (2 ten-millionths of a lux) to perceive it. That checks with our experience. We are able to see a full moon during twilight. In fact, we can see even a fraction of a full moon during full daylight.

In contrast, the North Star has an illuminance that is comparable to the threshold of human perception at the end of civil twilight. So the North Star is not likely to be perceptible until much later at night. In most large cities, with all the attendant light pollution, Polaris is not visible at all. But, free from light pollution, the skies of west Texas are among the darkest in the continental United States. This is one reason why the University of Texas located its McDonald Observatory in the Davis Mountains 38 miles north of Marfa.

After civil twilight, on a clear night with star cover and a full moon, we see from Table 5–1 that the threshold of human perception of light is about 4.0 E–08 lux (40 billionths of a lux). This level is substantially less than the illumination of Polaris. So, anyone on the Marfa plain — including the two campers in *Chapter 10 - Lights on the One O One Flat* — would have no difficulty seeing the North Star.

Knowing how much light must reach your eye in order for you to see it is one essential step in determining the distances over which lights can be seen. Another necessary ingredient is the strength of the light source — its intensity. Just intuitively, you know that a brighter light can be seen at a greater distance than a dim light. You might also guess that there must be some connection between the perception of your eyes, the strength of the light source, and the distance between your eyes and that source.

Well, there is a relation; and we don't have to invent it. We can follow the lead of those who have had a concern for seeing lights in the distance for many, many centuries — mariners. Seafarers have maintained light houses for millennia, and they have had ship lights and other navigation lights for centuries.

Today, these lights are regulated. The International Maritime Organization is concerned with the establishment of standards for marine activities. In 1972, they organized the development of an agreement — known as the "Convention on the International Regulations for Preventing Collisions at Sea" — that covers lights used in maritime activities. Today, nearly all countries have adopted the rules set forth in the convention, and these rules are written into the United States Code of Federal Regulations.[5]

The federal code includes an equation that specifies the intensity of a light in order for it to be visible at a prescribed distance. We will use this equation to determine the visibility distances associated with the lights on the Marfa plain.

The equation is:

$$I = 3.43 \times 10^6 \times T \times D^2 \times K^{-D}, \qquad \text{(Equation 5-1)}$$

where

I is light intensity in candelas,

T is visual threshold in lux,

D is distance between the light source and the observer in nautical miles, and

K is atmospheric transmissivity.

Wow! If you're not comfortable with math, you probably went into overload. Relax. The equation is for those who care. For those who don't, you can skip ahead to the paragraph designated by the **§**.

The federal code also specifies that the visual threshold (T) in the equation above should be 2.0 E–07 lux (2 ten millionths of a lux), and the atmospheric transmissivity (K) should be 0.8. These are the conditions of interest to seafarers — *i.e.*, visibility at sea level over a body of water where relative humidity is nearly 100%. Interestingly, the threshold (2.0 E–07 lux) is the same value we introduced in Table 5–1 as being applicable to the end of evening civil twilight.

Atmospheric transmissivity — a number between 0 and 1 — is a measure of how much light is ***not*** scattered out of a light beam. What does transmissivity mean to a person trying to see a light? Imagine that a beam of light begins propagating with a strength level of 100%, and one nautical mile later it is down to only 80%. Then the transmissivity (K) is 0.8. If 30% of the light gets scattered, K is 0.7, and so forth. So, low transmissivities mean you have to get closer to the light in order to see it.

Even though Marfa is far from major urban areas, and there is no heavy polluting industry, as we will see in *Chapter 10 – Lights on the One O One Flat,* there is often haze in the air on the Marfa plain.[6] Because of the common occurrence of haze, we used an atmospheric transmissivity (K) value of 0.8 in our calculations.

§ Visibility distances (D) for each manmade light source shown in Table 4–2 were calculated using Equation 5–1. We substituted values for the four human vision thresholds from Table 5–1 and the light intensity values (I) for the light sources into the equation. In this way, we calculated four possible visibility distances for each light source depending on the level of background illumination. The results are shown in Table 5–2.

The first column shows the light sources, while the second column gives the light intensities in units of candelas. (References for the intensity values are shown in brackets in column one.) Remember that intensity is a measure of the quantity of light projected into a solid angle in a particular direction. The more light that is projected, the greater is the distance at which the light can be seen.

Let's consider the intensity of the candle — the fourth item on the table. It has a value of 1.0, or one candela. This is no accident. Many years ago the quantity of light projected into a unit solid angle was defined in terms of a candle; and not surprisingly, this intensity was called *candle power*. The candle used in the definition of candle power had specific proportions, specific wick attributes, and specific wax composition. This special candle was the "standard" for light intensity, and its light intensity was defined to be unity. Today, *candle power* is an obsolete term. The modern unit of intensity is *candela*.

---

[6] See Reference [6] for a detailed U.S. government-sponsored study of pollution in the Big Bend region.

EMITTERS OF LIGHT	INTENSITY (candelas)	**HUMAN PERCEPTIBILITY** (statute miles)			
		Civil Twilight[3]	Full Moon	No Moon	Ultimate Limit
Campfire (small)	100	7	11	13	20
Match (burning after initial strike)	0.25	0.4	0.7	0.8	4.5
Match (peak light of strike)	2.7	1.9	3.6	4.6	9
Table candle [7]	1	1.2	2.4	3.2	7
Kerosene lamp used in home [8]	10	3	6	7	13
Train kerosene headlight [9]	9,000	20	25	28	37
Train carbon arc headlight [9]	81,000	27	34	36	46
Automobile kerosene headlights (averaged)	~2,000	15	20	22	31
Automobile incandescent headlights (averaged)[10]	~5,000	18	23	26	35
Electric light bulb (60-watt, frosted bulb) [11]	69	6	10	12	19
Mercury vapor lamp (100 watts) [12]	358	10	14	17	27

Table 5–2. Properties of lights available on the Marfa plain. Intensities came from references shown in column one. The distances at which the lights could be seen were calculated using Equation 5–1 shown above. For the calculations, the transmissivity of air was assumed to be 0.8. That level of transmissivity corresponds to a classification of the air as being between *clear* and *very clear*. Light sources are presented in the same chronological order as in Table 4–2.

Columns three through six show the four visibility distances for each light source. There is one distance for each of the background lighting conditions of Table 5–1. We found that a one-candela candle can be seen at civil twilight from 1.2 miles away. Under starlight and a full moon, the distance increases to 2.4 miles. Take away the moon (starlight only), and the distance increases to 3.2 miles. Finally, in a laboratory setting, the visibility distance more than doubles to 7 miles.

At first blush these results seem absurdly large. Take the match, for instance. Could you really detect a burning match from 0.4 miles away? We had to investigate that. Here's a simple test you can perform as well.

At the end of twilight (20 minutes after sunset), we positioned ourselves 0.4 miles apart and took turns striking wooden "kitchen" matches.[7] As a match ignited, we each observed an initially bright light flare. Then, the light dimmed somewhat as the match began to burn steadily. Finally, before the flame was completely extinguished, its light became too dim to be perceived by the viewer 0.4 miles away — confirming the prediction in Table 5–2.

While there is a rigorous definition of light output from the standard candle, there is none for a campfire. Because fire sizes and light outputs can range from a single burning twig to a raging forest fire, values for campfire intensity are just not to be found in the literature.

In order for us to provide something tangible for the visibility distance of a campfire, we set an estimate for the campfire intensity to 100 candelas. Why choose 100 candelas? Quite

---

[7] Hearthmark LLC, Diamond Brand, Extra Thick, "Strike on Box Matches".

simply, we reasoned that even a modest sized campfire could have a volume of flames that exceeded 100 times the volume of a candle flame.[8] Notice in Table 5–2 how the distances at which the campfire can be seen increase from those of the candle. Even under a full moon the 100-candela campfire could be seen from 11 miles away. (Obviously the fire needs to be high enough above the terrain, and there cannot be obstructions to the light.)

Now let's examine the kerosene lantern. Everyone knows that kerosene-fueled lanterns were used extensively in the late 19th and early 20th centuries in the U.S. But, you might be surprised to learn that these lanterns are still used extensively in the less developed countries today.

Researchers at the Lawrence Berkeley National Laboratory estimated that one out of four people in the world still obtains his light exclusively from kerosene and other fuels.[13] They also estimated that worldwide, fuel-based lighting costs $38 billion annually, and that fuel-burning lamps produce 190 million tons of carbon dioxide worldwide each year.

As a consequence, these lanterns have been studied extensively in the last two decades as part of formulating a global lighting strategy. Providing electricity to these users of lanterns would greatly reduce greenhouse gas emissions because one kerosene lantern produces 40 times as much greenhouse gas per unit of light as an ordinary incandescent light bulb!

Line three on Table 5–2 shows data for a typical kerosene-fired lantern. It is the kind of lantern with a wick that extends

---

8 In addition to the volume of flames, the *flame temperature* is a factor in determining the light output. We ignored the difference in flame temperatures of wood fires and candle flames.

down into a reservoir of kerosene. The light output from one of these lanterns is a modest 10 candelas. But the distances at which the lamp light can be seen are still large.

It is reported that when Robert Ellison got through the Paisano Pass and out onto the Marfa plain, he saw a light that he suspected came from a campfire.[14] But when he looked the next day, he couldn't find evidence of a fire or any other source of light. The entries in Table 5–2 for candle, campfire, and kerosene lamp all have visibility distances that would have permitted Ellison to see them on a starlit night (with or without the moon). And, the distances are too large for him to know for sure what he was seeing. Why couldn't he find evidence of fire the next day? Perhaps he just didn't ride far enough.

With the arrival of the Galveston, Harrisburg, and San Antonio Railroad in 1883, trains brought extremely strong sources of light — their kerosene-fueled headlights. Lines six and seven in Table 5–2 show data for train headlights measured in 1909. In that time period, train owners were being compelled by law to convert from kerosene headlights to carbon arc headlights. The Indiana Railroad Commission requested that Purdue University conduct tests to determine "the effect of such powerful headlights upon the interpretation of signals and upon the ability of the engineer to see and recognize obstacles upon the track in front of the locomotive".[9]

No wonder there was concern. The carbon arc headlight has the largest intensity of all the lights in the table with a value of 81,000 candelas. When carbon arc train headlights reached Marfa in the early 20th century, they could be seen from about 36 miles away — an almost incomprehensible distance. That's equivalent to the distance along US90 between Marfa and Valentine, Texas. If the earth were perfectly flat and there were no curves in the railroad tracks, the headlight of a train passing through Marfa could be seen in Valentine almost two hours before the train got there![9] Still, 81,000 candelas is small by today's standards. Now the U.S. Federal Regulations require that trains have headlights of at least 200,000 candelas.

As the 20th century began, cars and trucks joined trains in projecting bright lights on to the Marfa plain. There are no readily available records for the intensity of kerosene-fueled car lamps used in the early 1900s. But, we can estimate the strength of kerosene headlights by assuming that the lumens of a home kerosene lamp are all projected forward from a parabolic reflector into a 20°-wide beam. Two such lights on a car would provide about 2,000 candelas of ***average*** intensity and could be seen in civil twilight from 15 miles away.

In contrast to the kerosene headlight, the intensity of the incandescent bulb used in headlights, is a value that was measured at the time.[10] It was 21 candle power. In 1912, two headlights with 21 candle-power bulbs were capable of producing an average beam intensity that could be seen from 26 miles away on a moonless night. That's nearly as far as a carbon arc train headlight.[10]

No discussion of light sources would be complete without two modern lights. The ubiquitous 60-watt incandescent light bulb with an intensity of 69 candelas provides a basis of com-

---

9 See Footnote 3 in *Chapter 4 – Bringing Lights to Marfa* for a discussion of the speed of these early trains.

10 There was equal concern for the brilliance of automobile headlights. In 1918, the Governor of New York signed a bill regulating headlights. *The New York Times* heralded the bill — "It means the extinction of dazzling headlights, blinding instead of illuminative and conducive to accidents."[15]

parison that we're all familiar with. It can be seen from ten miles away on a moonlit night. You probably couldn't stand at the Marfa Lights Viewing Center (MLVC) and see light from one of these bulbs at a ranch house because of the vegetation and undulations in the terrain between you and the ranch. But you definitely can see mercury vapor lights (MVL) at the ranches — not only because of their strong intensity, but because they are mounted up on poles to illuminate the region around the buildings. Their height gets them above the vegetation on the plain that might obstruct their view from the MLVC.

Table 5–2 shows that a 100-watt MVL can be seen from ten miles away at twilight. And, visitors to the MLVC can confirm this for themselves. Looking toward the south just after the sun sets, you will see a very bright light become visible. It is at the Alta Vista Ranch about nine miles from the viewing center. There is a second light in the same general direction but somewhat higher up on the mountains south of the MLVC. That light only becomes visible after the Marfa plain becomes considerably darker. Even after you can distinguish it, you'll see the second light fade in and out. That light is about 17 miles distant — just what we'd expect from the visibility distance shown in Table 5–2.

Over the decades that people have been adding to the literature of the Marfa lights, the subject of visibility distances has come up repeatedly. Most often, it has been asserted that none of the known sources of light could have been what was seen because they were either too weak, or were too far away. We've just shown that common sources of light can, indeed, be seen from great distances, and this fact is important as we unravel the mystery that is the Marfa lights.

We'll encounter a variation of the "too far away" argument

later in this book. It's the assertion that "It couldn't be cars because there are no roads down there." But, if you go far enough on this planet, there is always "a road down there". As the map in Figure 2–1 shows, on the Marfa plain and its environs, there is a road in every direction, and not very far away.

Combine the high number of roads and ranches with the enormous distances that light can travel and still be perceptible to a person, and you have the opportunity for a lot of sightings. Compound the sightings with a belief that there is nothing nearby, add darkness, and the imagination takes over. Mythology is born.

—§—

5 — Sources

[1] Cecilia Thompson, *History of Marfa and Presidio County, Texas 1535-1946, Volume 1, 1535-1900* (Marfa, TX: The Presidio County Historical Commission, 1985), 194, 197.

[2] David L. Crawford, "Photometry Terminology and Units in Lighting and Astronomy," *The Observatory* 117 (February, 1997): 14-18.

[3] R.N. Clark, *Visual Astronomy of the Deep Sky* (Cambridge, England: Cambridge University Press and Sky Publishing, 1990), 11.

[4] W.E. Knowles Middleton, *Vision Through the Atmosphere* (Toronto: University of Toronto Press, 1952), 97.

[5] "Intensity of Lights," in *Code of Federal Regulations, Title 33, Subchapter E, Paragraph 84.15*. Available from: http://ecfr.gpoaccess.gov/cgi/t/text/text-idx?c=ecfr&sid=a6396ba406dff0936af39579b8b54acd&rgn=div8&view=text&node=33:1.0.1.5.39.0.26.8&idno=33

[6] Marc L. Pitchford and others, *Big Bend Regional Aerosol and Visibility and Observational Study [BRAVO], Final Report—September 2004*. Available from: http://vista.cira.colostate.edu/improve/Studies/BRAVO/Study

bravo.htm

[7] Nathaniel R. Greene, "Shedding Light on the Candela", *The Physics Teacher* 41 (October 2003): 409-414.

[8] Evan Mills, *Technical and Economic Performance Analysis of Kerosene Lamps and Alternative Approaches to Illumination in Developing Countries* (Berkeley, CA: Lawrence Berkeley National Laboratory, 2003).

[9] C. Francis Harding and A.N. Topping, "Headlight Tests," *Transactions of the American Institute of Electrical Engineers* 33 (May 17-December 31, 1910): 1053-1081.

[10] David W. Moore, "Headlight History and Harmonization," *Report No. UMTRI-98-21*, (Ann Arbor, MI: The University of Michigan Transportation Research Institute, June, 1998).

[11] General Electric Basic A19 Product Code 41026. Available from: http://genet.gelighting.com/LightProducts/Dispatcher?REQUEST=CONSUMERSPECPAGE&PRODUCTCODE=41026

[12] Westinghouse HF100R, Product Code 3740300. Available from: http://www.westinghouselighting.com/light-bulbs/hid-bulbs/mercury-vapor/100-watt-ed23-1-2-hid-mercury-vapor-light-bulb-3740300.aspx

[13] Evan Mills, "The Specter of Fuel-Based Lighting," *Science* 308 (2005): 1263-1264.

[14] Elton Miles, *Tales of the Big Bend*, (College Station, TX: Texas A&M University Press, 1976), 150.

[15] "Motor Headlights Must be Tested," *The New York Times*, May 12, 1918, Section 5, p. 3, col 1, Sunday edition.

6 — "Classic" Marfa Lights

RUTH BOWNDS WAS BORN IN MARFA in 1914 and lived there until she went away to college in 1932. During her teens, she, like others, saw mysterious lights in the open expanses both south and east of town. In particular, she saw the *classic* Marfa lights[1] south-southwest of the present Marfa Lights Viewing Center (MLVC).

What is really unique about her sightings is that Ruth de-

---

1 We describe these lights as the *classic* Marfa lights because this set of lights is the most documented. They are the ones for which the MLVC was built. There is even a tour bus that takes people from Marfa out to the MLVC to see these lights after sunset. But, there are other mysterious lights in the Big Bend region which are distinct from the *classic* Marfa lights.

scribed them in detail in her memoirs.[1] Writing about the classic Marfa lights, she said —

> *We . . . saw them later on Friday or Saturday night after a carload of kids had gone to the movies. We drove out the highway about 15 miles east of town, but not crossing the Paisano Pass uplift because they could only be seen from certain places. We pulled to the side of the road and looked down into a low place in front of a certain ridge. We could see them there parallel to the highway. The lights didn't move but looked like long streaks. Some thought they could have been automobile headlights, but there were no roads in that vicinity. So many people have walked down into that low arroyo looking for the source of the Marfa lights, but no car tracks or standing water which might have reflected light have ever been discovered. There has never been anything unusual about that dry area except the sighting of the Marfa Lights.*
>
> *Recently the Texas Highway Department has turned one of its roadside parks into a viewing area to see "The Marfa Lights". A small sign describes their history and directions for locating the lights. I had seen them in 1928-1932. They were still being seen in 1996. But, even now, there are no roads in that direction nor any explanations for the source of the "Marfa Lights".*[1]

What a marvelous description of a teenage memory! To our knowledge, it is the earliest account of the Marfa lights, to be written by someone who actually witnessed the lights. Ruth indicates that her classmates also saw the lights. But, if they wrote about them, we've not found any accounts available to the general public.

Ruth saw the classic Marfa lights from the Old Marfa–Alpine Highway east of town. She mentioned that the Texas Highway Department had built a viewing center where the lights could still be seen in 1996. Although we can't confirm that the MLVC was the exact site where Ruth's teenage sightings took place, we can say that she wasn't far from that location. Today, the lights can be seen along a fairly long stretch of US90 that extends roughly from Nopal Road to the curve of US90 leading to Paisano Pass.

Ruth doesn't indicate that the lights she saw were different from those seen from the Texas Highway Department's viewing area. So, it's probably safe to assume that the lights looked like stationary white spots of light just below the horizon. (In the next chapter, we'll address Ruth's reference to the lights appearing as "long streaks".) She also doesn't question the directions for locating the lights that are given by the sign at the viewing area. So, we can assume those directions are essentially correct.

With those facts established, we begin by laying out the area where the *classic* Marfa lights occur. The map in Figure 6–1 shows US67, US90, the Old Marfa–Presidio Road, and the terrain of the Marfa plain and mountains to the southwest. Chinati Peak — the highest mountain in the southern view — is further south than the region covered by this map but is quite visible from the MLVC.

The map also shows the MLVC on US90. To the east of it stands the entrance to the abandoned Marfa Army Air Field (MAAF). Built in 1942, the air field was operated until 1946. During World War II, thousands of pilots received their training at the MAAF and also contributed to the lore of the Marfa lights. Some of the ruins from the air field are still visible from the MLVC.

Note that in Figure 6–1, north is not at the top of the map.

The direction from bottom to top is approximately west-northwest. (See the arrow on the map.) The places on US67 where the lights can be seen lie between the two dashed black lines which intersect at the MLVC. The area between these dashed black lines, US67, and the MLVC has been lightened to highlight the viewing direction to the classic Marfa lights.

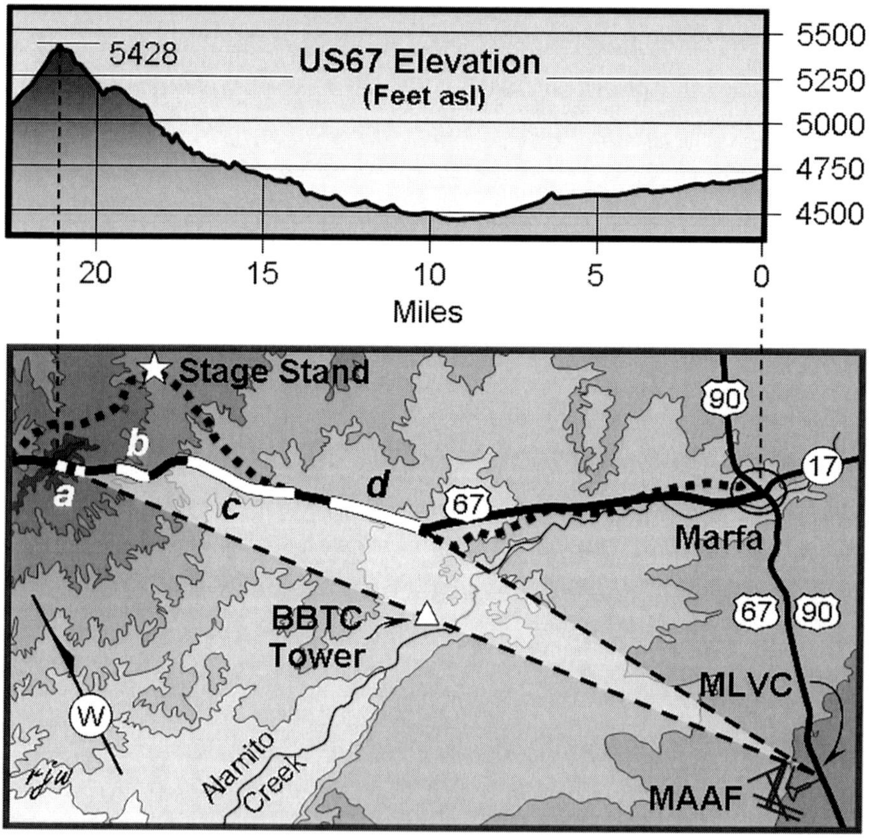

Figure 6–1. Illustration of features associated with the classic Marfa lights on the Marfa plain. The lower portion of the figure shows arrangement of features, roads, and topographic contours. The upper graph indicates the elevation along US67 from the high point of the terrain to Marfa, 22 miles away. The path of the old road connecting Marfa, Stage Stand, Shafter, and Presidio is indicated by the dotted black line. (Shafter and Presidio are off the left side of the map.)

From the MLVC, you can find the left side of the viewing area by locating the red light atop a Big Bend Telephone Company (BBTC) tower and then looking slightly westward. That tower is approximately 12.5 miles from the Viewing Center. The dashed black line from the MLVC to the BBTC tower continues on and intersects US67 in the upper left-hand corner of the map. The distance from the MLVC to that intersection is about 25 miles.

To find the intersection of the upper dashed black line and US67, a person at the MLVC first needs to locate Chinati Peak in the distance. The center of the segment labeled ***d***, the MLVC, and Chinati Peak are aligned with each other. Just to the right of that alignment is where the dashed black line defines the northern edge of the viewing area. This intersection of the dashed line and US67 is about 14 miles from the MLVC.

The lights occurring between the two dashed black lines are what people can see from the MLVC today. In print and conversation, they are known simply as *the Marfa lights* and are what we refer to in this book as the *classic Marfa lights* (CML).

In reality, these "mysterious" lights are the headlights[2,3] of vehicles traveling northeast along US67 toward Marfa.

Figure 6–1 also shows a dotted black line extending from Marfa southward. This line shows the path followed by the original road that connected Marfa, Stage Stand, Shafter, and

---

2 Everyone knows that vehicles have two headlights. But the classic Marfa lights are always reported as a single light. This will be explained in the next chapter.

3 See Reference [2] for an excellent analysis that proves the lights are produced by cars.

Presidio. Near Marfa, the old road meandered from side to side of the current US67, connecting the various ranches. It ran very close to the current US67 over the course of segment ***d***. But as it approached segment ***c***, it diverged westward following Perdiz Creek up the creek bed to Stage Stand at an elevation of 5,000 feet above sea level (asl). By following Perdiz Creek, and other creek beds beyond Stage Stand, the old road was able to cross the mountains between Marfa and Shafter without climbing to the peak of the range as the current US67 does.[4]

Vehicles have traveled this old route and scattered light across the Marfa plain since the late 19th century.[5] In the early 1900s, the Shannon Stage Line's coaches running between Marfa and Shafter used oil lamps for illumination. Then, in 1910, Shannon augmented its line with a five-person auto service.[3] By 1912, it was making daily round trips between Marfa and Shafter.[4]

With the rough road of the early 1900s and the number of creek crossings that needed to be executed, it is unlikely that the Shannon auto service's return trips to Marfa were always completed before sunset. Given the availability of electric headlamps as optional equipment in 1908 and standard equipment on some models in 1911 (see *Chapter 4 – Bringing Lights to Marfa*), the Shannon cars most likely had headlights. And, it is probably also safe to assume that Shannon cars were not the only ones in Marfa equipped with headlamps. So, there is little doubt that the road, that was the precursor to US67, was being used at night by automobiles at least as early as 1910.

---

[4] Today, US67 tops out at 5,428 feet asl. The old road reached an elevation of ~5,260 feet asl.

[5] The US Geological Survey shows this road on their 1895, 30' by 30' map of the Marfa region.

There is also no doubt that the auto lights that Ruth Bownds saw were on the Old Marfa–Presidio Road. Although this road was designated a federal highway in 1930, plans to improve the first 15 miles south of Marfa weren't approved by the Texas State Highway Department until April 1931, and bids for construction of the first 11 miles weren't awarded until March 1932. Since Ruth Bownds made her observations between 1928 and 1932, we know that she was not seeing cars on the current US67.

The fact that there have been headlights on the road between Presidio and Marfa since the early 20th century doesn't mean that Ruth Bownds and her friends saw anything like what we can see today from the MLVC. Not only were cars different back then, their headlights were nothing like those we have now. Later in this chapter, we will examine vintage car lights and what Ruth is likely to have seen in greater detail. But, first, we will discuss the lights you will see when you visit the MLVC today.

So, what's all the fuss about a few headlights in the distance?

Well, there is more to the CML than their identity suggests. For one thing, the lights are 14–25 miles away and are not recognizable as headlights. *And,* they do not *behave* as viewers expect. These lights grow brighter and dimmer; and appear to divide, merge, and separate. Most people would agree that's not "normal" headlight behavior. To understand what makes ordinary headlights "perform" mysteriously, we need to examine what viewers at the MLVC today are seeing.

Notice that in Figure 6–1, US67 is depicted in both black and white lines. The white segments indicate where the lights can be seen from the MLVC. Conversely, lights from vehicles on the portions of US67 that are indicated in black cannot be seen at the MLVC. Above the map is a plot of the elevation along the highway from the white segment labeled ***a*** all the way to Marfa. The plot is aligned with the map below.

The high point on the left side of the plot has an elevation of 5,428 feet asl. It is at this point that cars topping the crest have their headlights become visible to viewers at the MLVC and elsewhere on the Marfa plain.

At the other extreme — the end of white segment ***d*** nearest to Marfa — the lights disappear from view at the MLVC as the elevation of US67 becomes so low (4,480 feet asl) that higher terrain between the highway and the Viewing Center blocks all lights. The plot of US67 elevation shows that the road height declines about 990 feet from the high point shown on the left, to the lowest region before entering Marfa. The change in elevation is also visible on the contours of the map in the lower section of Figure 6–1.

During its descent, a vehicle meanders in front of hills, behind hills, down in small valleys, and up on crests of the terrain. It is this interplay between the terrain and the vehicles that causes the lights to appear and disappear. They disappear when higher terrain gets between the vehicle and the MLVC.

On the map, the segments where the lights are visible from the MLVC are labeled ***a***, ***b***, ***c***, and ***d***. The segments from ***a*** to ***d*** have lengths of 0.91, 1.0, 3.4, and 2.8 miles respectively. At 65 mph, a vehicle traveling from Presidio to Marfa would traverse segments ***a*** to ***d*** in 0.8, 0.9, 3.1, and 2.7 minutes, respectively.

The four white segments of US67 also have varying degrees of blockage by higher terrain between the highway and MLVC. Segment ***a*** has the most, but segments ***c*** and ***d*** also have a few obstructions that hide lights from view at the MLVC.

When a vehicle is on a portion of US67 that is aimed the most directly at the MLVC, the lights seen from the MLVC are the brightest. This is the case in the left-hand portion of segment ***c***. But about half way through that segment, the road angles more northerly, and the lights become dimmer. Conversely, when the road is directed strongly to the left or right away from the MLVC, the lights are not visible at all. This is the case between segments ***b*** and ***c***.

Segment ***a*** runs at an angle of about 14° from the direction to the MLVC, segment ***b*** is at ~10°, and segment ***d*** is at ~15°. Segment ***c*** has a bend in it. The southern portion makes an angle of ~4° with the direction to the MLVC, while the northern portion makes an angle of ~19°.

Ruth's description of the lights she saw didn't mention lights separating and then merging. Today, that action is frequently attributed to the CML. In fact, it is used by some as evidence that the lights cannot possibly be headlights. So, if the CML are really headlights, how do they divide and merge?

This "mysterious" behavior occurs when more than one vehicle is traveling on a segment of US67 visible from the MLVC. Let's consider segment ***c***, for example. Imagine that two cars traveling north on US67 have just entered segment ***c*** in Figure 6–1. The headlights of both cars are aimed almost directly at the MLVC.

The left side of the first cartoon panel in Figure 6–2 depicts this scenario. As they enter segment ***c***, initially the two cars' headlights are nearly aligned with one another. From 20 miles away, they appear to be a single light. When the leading car reaches the bend in the road, it begins to move out of alignment with the lights from the car behind it. After the lead car rounds the curve, it opens up an angular gap from the trailing car. Later, after the gap has developed, viewers at the MLVC see two lights where there had been one. Their perception is that the original, single light divided into two.

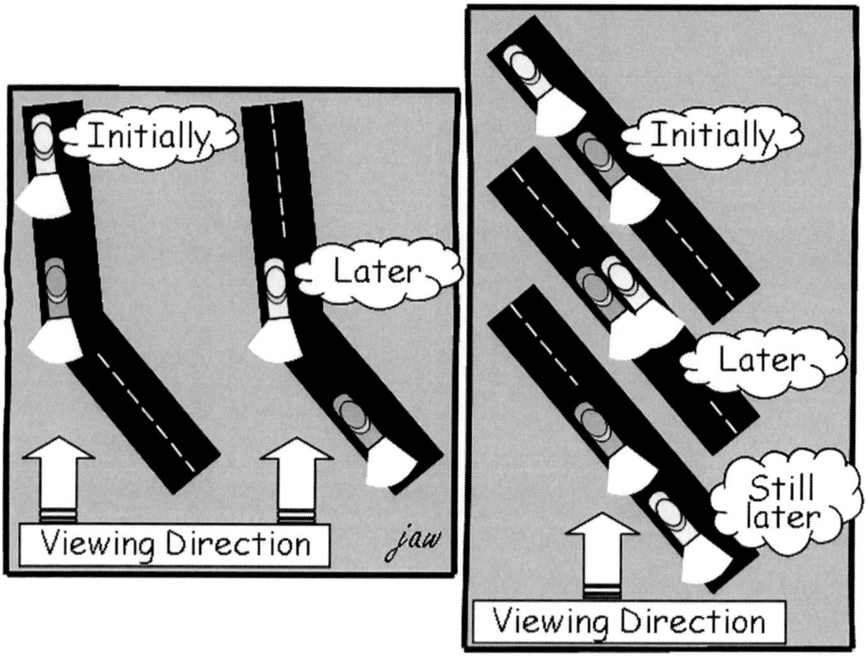

Figure 6–2. *Left side*: Positions of two cars at two different times illustrating how lights that initially appear to be a single light can divide when the first car reaches a curve in the road. *Right side*: Positions of two cars at three different times illustrating how two separate lights can appear to merge and then divide as one car overtakes and passes another car.

Another condition that leads to the perception of merging and dividing lights arises when one car overtakes and passes another. This might occur in segment ***d,*** for example.

Imagine that two cars enter segment ***d*** as shown in the upper portion of the second cartoon panel in Figure 6–2. Initially, there is distance between the cars, and their lights are seen as two distinct lights by viewers at the MLVC.

Now, let's assume that the trailing car is going faster than the leading one and later catches up with it. As shown in the middle of the right panel, when the two cars are side by side, a viewer 15 miles away perceives that their lights have merged into one. Still later, after passing the slower car, the faster car speeds ahead, opening up distance between them. As it does so, viewers at the MLVC perceive that the lights, which originally merged, have now divided — as shown at the bottom of the second cartoon panel in Figure 6–2.

Some may be disappointed at discovering that the CML are headlights and feel that knowledge detracts from the mystery. We aren't. Like the illusions created by magicians, the CML are captivating — even when you know "the trick". Where else but the Marfa plain can you see headlights perform so magically? The show is repeated every night at the MLVC and is fascinating to watch.

But, in contrast to what can be seen today, Ruth Bownds saw the lights from autos on the Old Marfa–Presidio Road when she made her observations between 1928 and 1932. It's not obvious that she saw anything like what we can see from the MLVC today.

For one thing, cars headlights of the late 1920s were substantially different from those we employ today. How do we know if the old autos were equipped with lights capable of projecting 25 miles across the Marfa plain?

As we saw in *Chapter 5 – Illuminating the Mystery*, a pair of electric headlights from the earliest automobiles emitted enough light (averaged over the beam) to be seen from 18 miles away **even in civil twilight!** After civil twilight, the visibility distance extended to 23 miles with a full moon, and on to 26 miles with no moon. From this data and the fact that Stage Stand (Figure 6–1) on the old road is about 23 miles from the current MLVC, we can conclude that in the early 1910s, automobile headlights on cars coming down the old road emitted just enough light to be seen by a person near the MLVC.

The nearest approach to the MLVC from which lights could be seen is shown by the upper dashed black line in Figure 6–1. At that point, the lights would be only 13 miles distant, and they would have shown brightly on viewers parked near the MLVC. But the early days' story of headlight visibility is more complex than this simple summary suggests.

The first automobile headlights were crude by today's standards. They were essentially spotlights. Spotlights are good for illuminating a small patch, but they are not good at providing a general awareness of the road both near and far ahead. For example, they might reveal that a bridge 200 feet in front of you is washed out, but not illuminate a boulder in the road ten feet ahead because the boulder is outside the light's narrow beam. To make matters worse, these narrow-beam lights were so intense that people and horses were momentarily blinded if the light shone directly into their faces. As a result of the difficulties associated with these "pencil" beams of light, experiments were conducted to reduce glare and broaden the beam.

Because of the variations in headlights during the early years of automobiles, we only quoted the "average headlight power" of the early lights in *Chapter 4 – Bringing Lights to Marfa*. The average power was much less than the *peak* power, and thus reduced the calculated distance at which the lights could be seen.

Table 6–1 shows an expanded set of facts about the headlights from automobiles in the 1911–1978 time period. The first line of data repeats the information shown in Table 5–2 in *Illuminating the Mystery*. It shows the average intensity over the entire ±10° beam of the headlight.

The second line shows what the same 1912 beam was reported to be capable of producing at its center — a staggering 30,000 candelas. If you were looking directly at that beam, you could see it from 24 miles away in the subdued lighting of civil twilight.

However, the problem with trying to see the headlight as the auto was traveling down the old road is that you'd need to be aligned with the direction of the beam to within a degree or so. That's a severe requirement. If you were not within that narrow range of alignment, you could not see the light. In fact, if you were looking from the location of the current MLVC, you couldn't see any persistent lights. But you might see a brief flash like that of a firefly every time the auto pointed straight toward you as it rocked up and down, and steered left and right negotiating the creek bed road.

All the wrangling over headlight intensity and beam shape taking place in the time period following the introduction of automobile headlights finally produced a standard for the performance of automobile headlights in 1922. (Before that time there wasn't a workable one.) This standard was one that made road illumination good enough for drivers, eliminated the blinding glare suffered by people looking toward automobiles, and still specified a headlight that could be manufactured with the technology of the time.

HEADLIGHTS	YEAR	INTENSITY (candelas)	**HUMAN PERCEPTIBILITY** (statute miles)			
			Civil Twilight	Full Moon	No Moon	Ultimate Limit
Average intensity over beam from 2 headlights[6]	1912	5,000	18	23	26	35
Possible peak intensity at beam center [5]	1912	30,000	24	30	33	42
Shaped-beam standard [6] — Peak intensity at 1° downward	1922	~10,000	20	26	28	40
Specification for peak beam intensity [7]	1940	75,000	27	33	36	46
Specification for peak beam intensity [8]	1978	150,000	30	36	39	49

Table 6–1. Evolution of automobile headlamp capabilities and the attendant visibilities of automobile lights. References are shown in brackets in column one.

The 1922 shaped-beam standard spread the intense central beam toward the sides, downward toward the region

[6] Average was produced by considering a 21-candela lamp with the lumens spread uniformly over ±10° beam.

closer to the automobile, and prohibited any strong light from being projected above the horizontal. That still left the headlight with a hefty 10,000 (or so) candelas shining about one degree down from the horizontal (and not in the face of an oncoming driver).

Since 1922, there have been two more updates to the headlight standard. The first in 1940 increased the maximum intensity to 75,000 candelas per vehicle. That coincided well with the development of the Marfa Army Airfield that brought thousands of World War II soldiers to the area to witness the mystery lights. With the increased intensity and increased width of the beam, viewing the lights was ensured. Finally, in 1978 the maximum intensity was increased to 150,000 candelas per vehicle. This is the currently allowed intensity in the United States and enables viewers to see lights along US67 from the MLVC.

While the 10,000 candela headlights of the 1922 vintage automobiles provided enough light to be seen from 26 miles with full moon (see Table 6–1), they still had a narrower beam width than is found on modern car headlights. That would restrict the places from which someone could see very distant lights. But Ruth Bownds says she saw the classic Marfa lights. And in 1928–1932, she probably was looking at headlights covered by the 1922 headlight standard. So how did she do it? Well, she had the benefit of an accommodating old road.

We stated above that the MLVC, the center of segment ***d*** in Figure 6–1, and Chinati Peak all lay along the same line. You can add Stage Stand to that list. Look back at Figure 6–1 and consider the sizable portion of the dotted black line that is aligned with the direction to the MLVC. That alignment is what enabled Ruth to see the lights in the 1928–1932 time period.

Traveling toward Marfa, the distance from Stage Stand to the place where the upper dashed black line on the map intersects the Old Marfa–Presidio Road is about 10.5 miles. From that intersection on closer to town, intervening terrain blocks the view of the old road from the MLVC. But for the 10.5 miles from Stage Stand to the dashed black line, about 70% of the time the old road points within 15° of the direction to the MLVC. In fact, if Ruth were parked closer to Marfa along US90 than the current MLVC, the alignment of these lights with her location would have been even greater.

That means the narrower-width headlight beams of 80 years ago were good enough for Ruth and friends to see them. And, because they were only visible when the autos were essentially pointing at the place where Ruth was parked, the lights appeared to be stationary — just as Ruth described. That also explains why Ruth didn't observe the lights dividing or merging. (There probably weren't that many cars traveling in close proximity to one another either.)

Ruth described the location where she and her friends parked as being about 15 miles east of Marfa and before Paisano Pass. To locate it, they *"looked down into a low place in front of a certain ridge".*

But, what landmark can be described as a "certain ridge"? Well, the Chinati Mountains do form the high portion of the horizon, and the end of segment ***d*** on US67 is near the low point on the Old Marfa–Presidio Road. So, segment ***d*** could be described as being in a "low place in front of a certain ridge".

With this interpretation, it seems that Ruth was talking generally about lights from the direction of segment ***d*** on US67. Indeed, the old road did lie within that angular range. It's a very much narrower angular range than what can be seen by viewers of the lights on US67 today.

Ruth said, "So many people have walked down into that low arroyo looking for the source of the Marfa lights, but no car tracks or standing water which might have reflected light have ever been discovered."

Why didn't they find anything in the arroyo?

Well, the direction they were looking (along the upper black dashed line in Figure 6–1) passes right over the top of the union of Long Draw and Colquitts Draw. That junction is about four miles from the MLVC and is the low point in the nearby terrain. If Ruth and friends thought that the small dots of light were close by, then this "low arroyo" would seem to be a logical location for the mysterious lights they were seeing. Little did Ruth and her friends imagine that the lights they were observing were actually between 13 and 23 miles away!

These classic Marfa lights weren't the only mysterious lights Ruth Bownds observed as a teenager. In the next chapter, *Lights Over the Alamito*, we'll investigate her sightings from the Jordan Ranch, discuss those light streaks she talked about, and explain how automobile headlights appear to be a single light, rather than a pair.

—§—

6 — Sources

[1] Deric Bownds. "Marfa Stories told by Ruth Bownds Kershner", January 8, 1999. (Compiled by Susan Kershner Mundine, November 2001.) Available from: http://www.dericbownds.net/Bownds_History/Marfa_Stories/

MarfaStories.htm

[2] Society of Physics Students, "An Experimental Analysis of the Marfa Lights," December 10, 2005, University of Texas at Dallas. Available from:
http://www.spsnational.org/wormhole/utd_sps_report.pdf
[The link in the PDF to the videos they produced no longer works.]

[3] Cecilia Thompson, *History of Marfa and Presidio County, Texas 1535-1946, Volume 2, 1901-1946* (Marfa, TX: The Presidio County Historical Commission, 1985), 73.

[4] Thompson, 92-93.

[5] David W. Moore, "Headlamp History and Harmonization", *Report No. UMTRI-98-21.* (Ann Arbor, MI: The University of Michigan Transportation Research Institute, June 1998), 2.

[6] Lighting and Illumination Committee, "State Regulation of Motor Car Headlamps," *Journal of the American Institute of Electrical Engineers* 41, no. 3 (March 1922): 223-225.

[7] Moore, 10.

[8] Moore, 13.

7 — Lights Across the Alamito

IN THE LAST CHAPTER, we examined the lights Ruth Bownds saw as a teenager when she and her friends parked east of Marfa at night. The sources were unidentifiable to them but the lights were always visible from a location along US90 near the current Marfa Lights Viewing Center (MLVC).

These were the *classic Marfa lights.* And, their mysterious sources were the headlights of cars heading toward Marfa on the Old Marfa–Presidio Road. Today, visitors to the MLVC can witness the *classic Marfa lights* of US67 as they appear and disappear along the southwestern horizon.

But, Ruth also mentioned seeing other lights that were not in the area of the *classic Marfa lights*. These lights were east of

Marfa.

> *In 1928, when I was 14 and in Marfa High School, dancing and card playing in this Baptist community were equally sinful. . . .*
>
> *One of my classmates, Kathryn Jordan, lived in a big two story house on the south side of town, no other roads or houses beyond it. Although most Baptist parents considered "couple" dancing sinful, her parents, who were also Baptist, didn't mind couple dancing. Kathryn's mother, Kate Jordan, organized private parties for the teenagers and offered space for dancing. We helped her sweep the hay out of the barn so we could dance in the evening. The boys would never ask a girl to dance and couldn't really dance, but they seemed to enjoy the group activity. At night, looking east over empty pastures, we saw the mysterious Marfa Lights each evening.*[1]

Now, that's just about as explicit as it can be. If you want to see the lights, look east from Kathryn Jordan's barn. But where's the barn?

These lights were seen about 80 years ago. Barns can disappear over the course of time. Indeed, this one has too. Ruth described the house as two-story. There aren't very many two-story houses in Marfa. But driving around looking for a two-story house that fits Ruth's description isn't an effective research technique. We tried the phone book — hoping to find a Jordan who might fill in the history. No luck there either. Clearly, finding the site of the Jordan barn was going to require some detective work.

Who *were* the Jordans?

Like Ruth Bownds, her friend, Kathryn Jordan, was also born in Marfa in March of 1914. She was the daughter of

Forrest W. and Kate Russell Jordan. Forrest was bookkeeper for the Murphy and Walker Company in Marfa. Over the years, he bought and sold a few properties in town. The one we care about was acquired in June, 1929 from F.A. Mitchell. It was slightly larger than 2 acres and located just south of the Marfa town limits. The deed described the property as:

> *. . . being situate* [sic] *in Presidio County, Texas, being a subdivision of Survey No. 241, Block 8. G.H. & S.A. Ry. Co., described by metes and bounds as follows, to wit:*
>
> *Beginning at a point, stake for corner, where the center of what is known as Davis Street in the Town of Marfa, Texas, intersects a fence bordering the Marfa–Alpine Road, said fence being sixty feet South of the South line of Section 244, Block 8. G.H. & S.A. said S line of said Section 244 being also the N line of said Section 241; Thence South 300 feet to stake and S.E. Corner this tract; Thence West 320 feet to stake for SW corner this tract; Thence North 300 feet to stake for NW corner this tract; Thence East 320 feet to the place of beginning.*[1][2]

This description may seem impossible to locate; but, at least, it didn't use old oak trees or boulders to define the property corners. That was a break.

In deciphering the description, we encountered two surprises — reference to Davis Street and reference to the Marfa–Alpine Road. Look on any current map of Marfa, and you won't find a Davis Street, nor mention of the Marfa–Alpine Road.

No one at the Presidio County Clerk's Office had heard of Davis Street either. But, they permit investigators to go

1 Spelling, abbreviations, capitalizations, and spacing have been faithfully reproduced from the original deed. The warranty deed is on file at the Presidio County Clerk's Office, Marfa, Texas.[2]

through the records stored in their massive vault. While looking at old maps, we found it! A map of Marfa streets dated 1884 with Davis Street! Davis Street was the original name of Highland Street. It ran right up to the courthouse square!

But the old map didn't yield any clues to the location of the Marfa–Alpine Road. Confirmation of the road's path was finally obtained from the detailed construction drawings for US90, approved by the State Highway Department in 1931. The drawings also provided another surprise. The old road to Alpine from Marfa did not follow the current path of US90 out of Marfa.

Originally, the Marfa–Alpine Road followed the path of the current Madrid Street at the southern end of Marfa. Proceeding eastward, it crossed Davis Street (currently Highland) and Alamito Creek. East of the creek, the road traveled over the open fields to approximately location ***J*** on the map in Figure 7-1. From there, it followed the path of the current US90, and topped the hill crest on the right side of the map. Today, the junction ***J*** is just south of several large, white telecommunications sheds located between the Southern Pacific Railroad tracks and US90.

The Marfa–Alpine Road was maintained by the state. Lest anyone think state maintenance indicated a high standard for the roadway, road maps of the time suggest otherwise. The section of road between Marfa and Alpine was classified among "improved graded roads" on the Automobile Red Book's 1928 Map of Texas, but only "semi-improved [impassable-wet weather]" on the 1931 E.L. Steck Company's Highway Map of Texas.

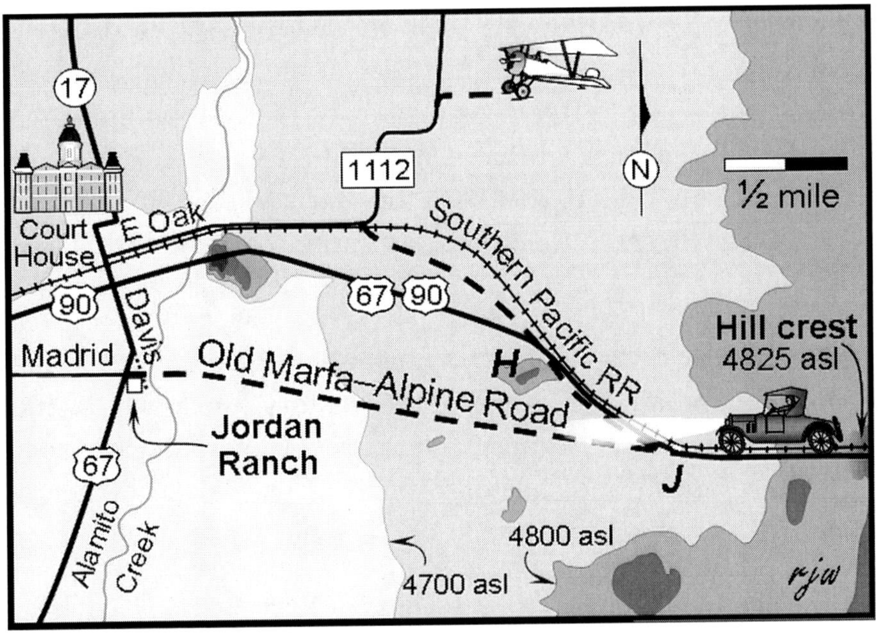

Figure 7–1. Relation of the Jordan Ranch (white square) to the Old Marfa–Alpine Road and the current roads and railroad in the region. (The old road to Presidio that would become US67 was off the left side of the map. It did not run past the Jordan property as US67 does today. See map in Figure 6–1.)

The Marfa–Alpine Road and the other commonly traveled paths of the 1920s and 1930s are indicated by the dashed black lines on the map in Figure 7–1, while the modern roads and railroad are designated by solid lines.

Another landmark of the time is the airfield located in the upper portion of Figure 7–1. This facility — built in 1919 to support the U.S. Border Patrol efforts — required a road connection into Marfa. Aerial photographs from 1919 show a well-traveled road that entered Marfa at East Oak Street. Today, Texas RM1112 follows the same path out of town to the Marfa Golf Course.

With Davis Street and the Marfa–Alpine Road identified,

we were finally able to locate the original Jordan property. Designated by a white square in Figure 7–1, the ranch sat south of the current Madrid Street — outside the Marfa town limits — just east of present day US67,[2] and right on the side of Alamito Creek. In the period 1928–1932, the Marfa–Alpine Road intersected Davis Street at the northeast corner of the Jordan property.

Today, partially hidden by more modern buildings along US67 and nestled among trees, a two-story house sits in the location defined by the deed. The barn is gone; and whether this house is the one the Jordan's lived in, we've not determined.

At last, we had the site for viewing Ruth Bownds' eastern lights. From the Jordan's place, you could look eastward, more or less directly along the Marfa–Alpine Road. The view is unobstructed for about three miles until a hill crest about 150 feet higher in elevation than the Jordan Ranch gets in the way. The bearing of the highest portion of the road — from the crest downhill toward the Jordan Ranch — lines up with the viewing direction to within a few degrees.

Looking at the map, you'll see the course of US90 from location ***J*** to San Antonio Street in Marfa. This road didn't exist in 1928–1932[3] — probably because it required a substantial cut

[2] The Presidio–Marfa Road (shown on 1920s maps as State Highway 17) was designated US Highway 67 (US67) in April, 1930. It entered Marfa to the west of the current US67.

[3] Approval for construction of US90 in Presidio County — signed November 20, 1931 — called for the highway to enter Marfa at San Antonio Street. Construction on US90 between Alpine and Marfa began in March, 1932. The Old Marfa–Alpine Road that entered the south

through the hill just outside Marfa. But that doesn't mean there was no traffic in this area.

It is likely that much of the traffic traveling from the east to Marfa turned off the official Marfa–Alpine Road at ***J*** and simply drove through the fields, crossed the creek bed, and entered Marfa on the south side of the train tracks — avoiding the longer distance from ***J*** along the Old Marfa–Alpine Road and then north on Davis Street. Common paths developed south of the railroad tracks and east of Marfa. They are visible on aerial photographs as early as 1919. But, they amounted to no more than trails through the fields.

One path that is particularly relevant to this discussion is the one indicated by the dashed black line from location ***J*** northwest across the railroad tracks where the path joined the airfield road. (Today Golf Course Road off US90 follows the same path north.)

Just as travelers to the south side of Marfa left the highway at ***J*** and drove directly to the desired city street, those going to the north side of town would have followed the dashed line across the railroad tracks to reach the airfield road, entering town on Oak Street. Still others would have followed alongside the south side of the railroad tracks to Alamito Creek and driven down the bank and under the railroad trestle to reach the north side of the tracks. (People still drive along the Alamito Creek bed and under the railroad trestle today.)

The lights Ruth Bownds and Kathryn Jordan saw 80 years ago first became visible as cars topped the crest of the hill about three miles away. Identifying them as car headlights

---

side of town was maintained as a detour during construction and then abandoned.

would have been difficult at that distance. If the lights had remained visible right up to the Jordan Ranch, it would have become obvious that they came from cars.

Since Ruth does not correlate these eastern lights to cars, we have to conclude that they remained in the distance and did not pass the Jordan property. Most likely, the vehicles took a short cut along one of the trails into Marfa, and as the route deviated from the Marfa–Alpine Road, the headlights simply disappeared from view. Remember, these were 1922-vintage headlights with a narrow beam width. So when vehicles weren't aimed at the Jordan property, the headlights "went out".

Visitors to Marfa can see approximately the view Ruth saw from the Jordan Ranch. Just go south on Highland Street 0.6 miles from the Paisano Hotel to the intersection of US67 and Madrid Street and look east along the Old Marfa–Alpine Road, currently a private drive to the MacGuire Ranch.

The crest of the hill is visible in the distance. When a car coming toward Marfa reaches the crest on US90, its headlights become visible to viewers at Madrid and US67. At this point, the car is only about three miles away. The lights flare in brightness as the angle of the beams comes into alignment with the viewer's gaze. The lights then seem to weaken but remain visible until the automobile rounds the curve at ***J***, and disappears from view as it passes behind the hill shown at ***H***. When it is lost from view, the car is 1.9 miles away.

Today, cars — equipped with modern headlights — are traveling 70 mph along this route. Being so close to the intersection of Madrid and US67, it's obvious to ***us*** that the lights are from cars. But in 1928, the cars were bouncing along an "improved graded or semi-improved" dirt road — the Old Marfa–Alpine Road — at speeds more likely to be around 30 mph and shining headlights with only 13% the intensity of cars today. Like the *classic Marfa lights* Ruth described, they weren't distinguishable as headlights.

Vegetation and buildings somewhat limit your ability to see the car lights from the intersection by the Jordan Ranch property today. And, you can't go east on the Old Marfa–Alpine Road because east of US67, Madrid Street becomes a private drive leading to the MacGuire Ranch.

But there is another location for viewing these eastern lights. Above the old Camp Marfa, the Pinto Canyon Road (RM2810) intersects with Caa Street. At the intersection of Caa and Yale Streets, you can see to the eastern ridge the way Ruth did in 1928.

This location is only 37 feet higher in elevation than the old Jordan Ranch, but it is above the local obstructions that limit visibility down lower. From the eastern end of Caa Street, the lights are very clear. This is the same sort of display that is visible from the MLVC looking toward US67, but in this case, the viewing location is only about one mile from the center of Marfa.

What you notice immediately is that you can't tell whether the automobiles have two headlights or one. The question of associating distant lights with an automobile raises the specific question of how far away a person can be from two headlights and still be able to distinguish them as two *separate* lights and not a single, large source of light. Well, that's a question that can be answered with confidence. There are two factors that determine the answer — the physics of light propagation, and the physical limitations of the human eye.

The physics, and related mathematical analysis, of circular optical instruments was conducted in the nineteenth century by Lord Rayleigh. The answer is that the resolution cannot be better than about 1/3 of a minute of arc. That's 1/180th of a degree, or two lights that are separated by 4.5 feet and 8.8 miles away. Certainly, that would be good enough for lights seen from the Jordan Ranch along the Old Marfa–Alpine Road.

Unfortunately, optical instruments can be made with a perfection that is not matched by the human eye. A person with excellent vision might only be able to resolve two objects, ***in the daylight***, that are 1.2 minutes of arc separated from one another. It gets worse at night. In the dark, when the pupils of a person's eyes open wide to collect more light, visual resolution is reduced by nearly a factor of three from daytime visual acuity.[4] Under the condition of fully open pupils, car lights that are 4.5 feet apart cannot be distinguished from a single light at distances greater than about one mile.

Try viewing the lights east of US67 and Madrid Street (or better yet from Caa and Yale Streets) with and without your glasses and see what you perceive. The authors have tried it. With and without glasses, we were only able to distinguish a single light.

Aberrations of the eye lenses produce other visual effects as well. Ruth Bownds described the distant (13–23 miles) *clas-*

---

[4] Lens distortions in the human eye are larger near the outer edges than they are near the center of the lens. At night, the pupil opens wide to permit more light to enter the eye. But this action also enables greater distortion of the image because light passes through the less perfect outer edges of the lens.

sic Marfa lights as streaks, not points of light. Seeing streaks, instead of points, of light was most likely the consequence of irregularities in the lenses of her eyes — perhaps the result of astigmatism.

The Marfa lights we've discussed thus far have been headlights that appeared to be mysterious because the observer couldn't see enough detail (two lights versus one) to identify them as headlights and the setting was too distant to put the lights into a familiar context. So, the observer couldn't tell the lights were following a road, nor see the interplay of multiple cars, nor identify that obstructions of terrain interfered with the lights, causing them to "turn off and on". Without these facts to anchor the viewer's thoughts, anything can be imagined. And, ordinary lights become extraordinary in the pitch-black darkness of the Marfa plain.

But there is more to the Marfa lights, and the distortions which can be applied to them, than we've discovered thus far. When the "forces of nature" act upon lights, the effects can be truly mysterious. In the following chapters, we'll examine how the atmosphere and meteorological conditions can produce lights that do the unexpected.

—§—

7 — Sources

[1] Deric Bownds. "Marfa Stories told by Ruth Bownds Kershner", January 8, 1999. (Compiled by Susan Kershner Mundine, November 2001.) Available from: http://www.dericbownds.net/Bownds_History/Marfa_Stories/MarfaStories.htm

[2] Warranty deed. Grantee: FW Jordan; Grantor: Mitchell, FA *et ux*; Transaction date: 06-01-1929; Record date: 03-01-1930. *Deed Book 87.* Marfa, TX: Presidio County Clerk's Office, pp. 162-163.

8 — Bending Light Beams

THE WORD *MIRAGE* conjures up images of parched prospectors crawling through the desert toward an elusive oasis, or mariners seeing a ghost ship on the sea's horizon. But, mirages are not the illusions of fiction. They are created by real phenomena. You see them as pools of water that appear and "evaporate" before your eyes as you drive down long, straight stretches of highway. Or, more dramatically, they can appear to swerve across the highway, as the one on the front cover of this book does.

If mirages aren't the product of delirium, what are they? And, what do they have in common with mysterious lights? The answers are found in the *bending of light beams*.

To appreciate how mirages are produced and what makes the Marfa lights appear to be mysterious, we need to examine what happens as light propagates from its source, through the air to our eyes. At first thought, the explanation seems simple enough. In elementary physics courses, we're taught that light travels in straight lines.

Throughout most of the universe, this is true. But most of the universe is essentially a vacuum, devoid of matter. It is also a homogeneous medium. A homogeneous medium is simply one in which the properties of the medium do not change with position in the medium. It is this characteristic that enables straight-line propagation of light.

For example, consider a swimming pool filled with water. The optical properties of the water are the same everywhere in the pool. So, light travels in straight lines in the water. The same is true for the air in a room in your house. With the doors and windows closed and the air temperature outside the house about equal to the air temperature inside the house, the air in your room is homogeneous for all practical purposes; and thus light travels in straight lines in the room.

On earth, the light we see usually travels from its source to our eyes through the atmosphere — an atmosphere that is ***not*** homogeneous. That inhomogeneous environment is very different from traveling in the vacuum of outer space. The inhomogeneity of the earth's atmosphere generally causes light to travel in slightly curved arcs through the air of our planet.

Some of the inhomogeneity comes from composition changes, but most of it comes from temperature and pressure differences. To illustrate, let's consider two locations about six miles apart. The first location is on the earth's surface near sea level. The second location is directly above the first at an altitude of six miles. At the first location, the air pressure is that of the full atmosphere — a pressure imaginatively termed, *one atmosphere*. At location two, the air pressure is only about 25%, and the density about 35%, of their respective values on the surface of the earth. In addition to the pressure and density differences, the air temperature six miles up is likely to be 120°F lower than the temperature on the surface of the earth.

The differences in temperature and pressure between the two locations are big. These big differences are achieved by a succession of small differences in the variables along the six-mile descent through the atmosphere. Each small difference in air temperature and pressure from one altitude to the next produces only a little bending of a light beam. But the cumulative effect of these successive small differences can strongly bend light beams[1] that are propagating somewhat parallel to the surface of the earth.

Imagine you're the observer on the earth shown in Figure 8–1. You're standing outside observing the stars, or perhaps watching a sunset. Light comes from an astronomical object off to the left. It travels through the vacuum of outer space in a straight line. As the light approaches the earth, it first enters the atmosphere about 20 miles above the surface and 400 miles away from you.

Notice that the light starts to curve toward you. As the light propagates through the air, it is always bent in the direction of the denser air. That means the light bends downward, toward

1 Strictly speaking, the ***path*** traveled by the light is being "bent", not the light itself. But throughout this book, we'll use the shorthand language of "bending light beams".

the surface of the earth where you're standing. If you're watching a sunset, the rays of the sun would be coming in tangent to the earth's surface, and a strong bending of about half a degree would occur. If you were looking more upward through the atmosphere at a star, a smaller amount of bending would occur. And, if you were looking straight up, the light beam would not be bent at all.

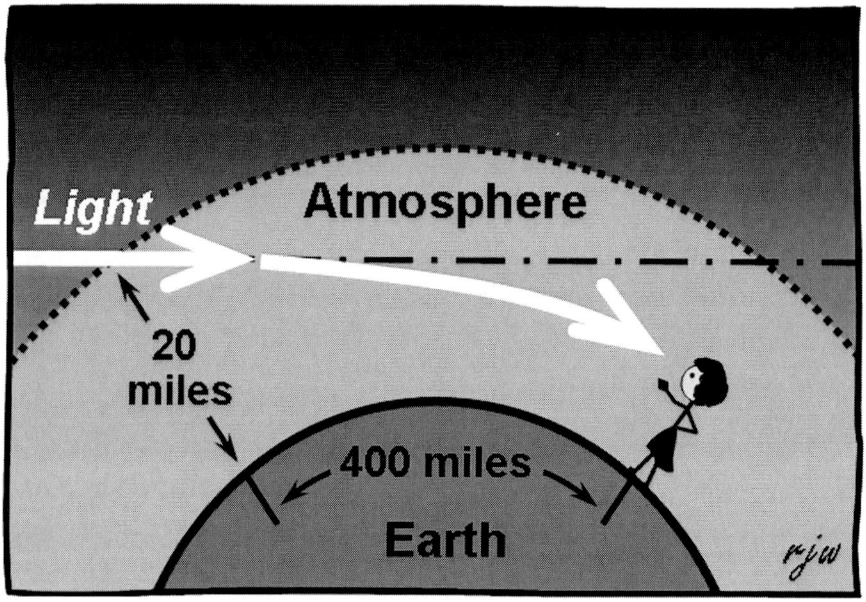

Figure 8–1. Illustration of light from an astronomical source being bent by the earth's atmosphere downward to the observer's viewing position. If the observer were at sea level, and looking out horizontally at the last rays of a sunset, she'd be seeing light rays that were bent about a half degree by the atmosphere.

The bending we're talking about is called ***refraction***, from the Latin, meaning *to break*, or *force back*. And, as we'll see, refraction plays a role in the creation of mirages and the propagation of mysterious lights on the Marfa plain.

Refraction of light coming through the earth's atmosphere is not a recently discovered effect. Nor is it a product of NASA and the space age. Refraction has been known for centuries, and the magnitude of the bending has been tabulated in textbooks on astronomy and surveying.

In 1695, Sir Isaac Newton calculated a table of refraction values for light traveling through the earth's atmosphere.[1] And, in 1723, James Hodgson qualitatively discussed how refraction by the earth's atmosphere altered the apparent positions of astronomical sources of light. He provided instructions for correcting observers' readings of angles.[2]

By 1821, Nathaniel Bowditch had already released the fifth edition of a text with comprehensive tables that enabled mariners to correct their sun and star readings while at sea.[3] And, in the early 20th century, refraction-correction procedures were commonplace in texts on surveying.[4]

Surveyors have an added problem. They have to account for refraction as well as for the fact that the earth is curved. In his surveying book, Henry Adams included rules-of-thumb for correcting for the earth's curvature and for refraction of light over a path between a particular land object and the surveyor.[4] The refraction correction is one-seventh the magnitude of the correction needed for curvature of the earth over the same distance.

Let's illustrate these corrections. Imagine you are looking at a mountainside 20 miles away. That's about the distance from the Marfa Lights Viewing Center (MLVC) to the mountains south of the Marfa plain. In order to get the true elevation above sea level (asl) for a location on the side of a mountain, you need to adjust for the curvature of the earth.[2] Henry Adams' rule-of-thumb says that an object 20 miles away is about +270 feet higher above sea level than it would be on a flat earth. Now, to account for daytime refraction of light by the atmosphere, you need to subtract about 38 feet — one-seventh of the correction that was used for earth curvature. Thirty-eight feet over 20 miles may not seem like much, but there are other factors we need to consider in deciding whether 38 feet is significant.

First, even 38 feet of bending of a light ray puts it on a scale with the shrubs and hillocks on the Marfa plain. Secondly, refraction during the daytime is substantially smaller than refraction that occurs at night on the Marfa plain, as we'll see in *Chapter 9 – Larger Than Life*. If you applied daytime refraction rules for some nighttime conditions, you might get misleading results. For example, if you were to set up a camera during the daytime with it aimed such that the roads below a certain angle were not visible (as was done in Reference [5]) those roads might become visible to the camera at night due to the stronger bending of light rays.

To explain how refraction alters our perception of an image, and how it creates mirages, we need to examine the physics underlying refraction. Mindful that not everyone is mathematically or scientifically inclined, we'll keep the discussion as simple as possible and illustrate the effects with examples and pictures. Not only will the mechanics of refraction be revealed, but in the process, two rules will emerge that enable us to ex-

---

[2] Curvature of the earth is not something most of us think much about. That's because we're used to seeing things that are near by. But in the open spaces of the desert southwest, you can see for tens of miles; and distant objects tend to disappear over the horizon. That sinking in the distance is due to the curvature of the earth.

plain how mirages work.

During our discussion of Figure 8–1, we said that light is bent in the direction of higher atmospheric density. More fundamentally, the light is bent in the direction of increasing ***index of refraction***.

All media through which light propagates are characterized by a parameter called the *index of refraction* (IOR). The IOR of air is affected significantly by five factors. The first of these is a property of light itself — its color. So, the IORs of air for red light, yellow light, and blue light are all different from one another. The other four factors affecting air's IOR are all properties of the air — its temperature, pressure, relative humidity, and carbon dioxide content. The relationship of the IOR to these factors is exceedingly complex, and accurate **theoretical** formulas do not exist.

Light traveling from a region with one value of IOR into a region with a different IOR is bent at the junction between the two media. It is rigorously correct to attribute the bending of the light beam to the difference in the two IORs. In the case of atmospheric air, it is also correct to speak of the bending as being caused by the difference in the air density of the two media because higher atmospheric density also produces higher IOR. Since density is a more familiar concept than IOR, we'll commonly speak of the air density in the discussions below.

Consider the illustration of light propagating from upper left downward to the observer in the lower right of Figure 8–2. The darker lower region is denser and has the larger IOR of the two layers. For simplicity, we'll assume that there is an abrupt transition from the top layer to the bottom layer at the horizontal interface between the two.[3] Notice that the light is bent downward *(refracted)* at the transition into the lower layer of air.

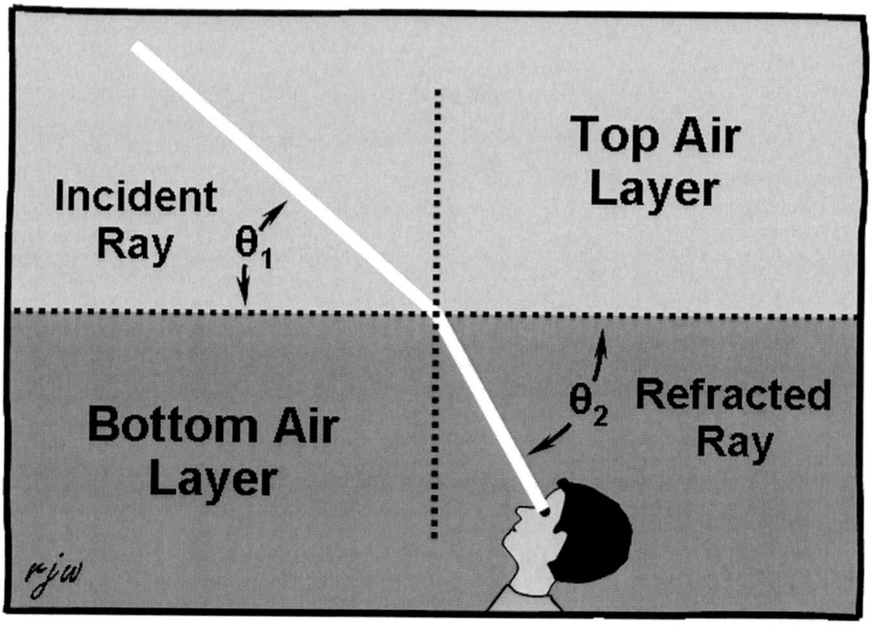

Figure 8–2. Light propagating between two layers of air that have different IORs. The darker bottom layer indicates denser air with the larger IOR. Whether the light travels downward or upward, its path is bent at the interface toward the air with the larger IOR.

There is a mathematical relation involving the IORs and the angles of the light rays from the interface between the two layers. It is known as Snell's Law and was described by Willebrord Snell in 1621 (see *Glossary*). Snell's Law is applicable to any two optically transparent media. So air/air, air/water, air/glass, water/glass, etc. are all described by the law. Snell's simple equation tells you what angle (θ_2) the refracted ray will

---

[3] It can be shown that even if the change from the top IOR to the lower IOR occurs over a sequence of small steps instead of one large step, the amount of bending is the same.

follow after the incident ray strikes the interface.

Now, instead of projecting the light from the top downward, let's shine the light in the opposite direction. Imagine that the light ray is traveling from the position of the person in Figure 8–2 ***upward*** through the atmosphere. At the interface between the two layers, the light once again bends downward *(refracts)* toward the layer with the larger IOR — the denser bottom layer. ***It is a general truth that in air, a light ray bends toward the denser medium regardless of the direction in which the light is traveling.***

As we said, Snell's Law is applicable to any two layers of optically transparent media. But in order to make a specific calculation of the bending, we need numerical values for the IORs of the two media. For many materials, these numbers are tabulated in handbooks. But, in the atmosphere, in which the IOR changes so slightly from one height to the next, more accuracy is needed.

The United States National Institute of Standards and Technology (NIST) has provided formulas for the IOR of the earth's atmosphere on their web page.[6] The documentation and equations available there enable someone to calculate the IOR of the earth's atmosphere to an accuracy of better than one part in 10 million. We'll use NIST's formulas in all our calculations of light propagation in this book.

Let's examine a specific condition that you might find around Marfa in the summer time. Visualize the air over an asphalt road. At the height of a person's eyes standing on the road (or the height of a person in a car driving on the road), the air temperature is 90°F. But down at the asphalt surface, the temperature is 140°F. And the air in contact with the asphalt is also 140°F. Within the first one-half inch or so above the asphalt, the air temperature quickly makes a transition from 140°F to 90°F. So, let's approximate the transition as occurring abruptly at one-half inch up from the road. Now, we have two distinct layers of air with a sharp interface between them.

At 90°F, we used the NIST equations to calculate an IOR of

$$\text{IOR}(90) = 1.00022395$$

and at 140°F, we got an IOR of

$$\text{IOR}(140) = 1.00020429.$$

The IORs for both temperatures of air are very close to unity. While the IOR of the 90°F air is greater than that of the 140°F air, the two values do not differ very much from one another. But even this small difference in IOR will cause a light beam to bend.

The amount the light beam will bend is affected not only by the IORs, but also by the angle at which the light is incident, *i.e.*, the angle of the input ray relative to the interface between the two layers.

The two-layered arrangement of air and asphalt is shown in Figure 8–3. As before, the darker layer is denser than the lighter layer. But in this case, the dense layer is on top. The bottom layer is very thin, approximately one-half inch thick. Below this very hot, thin layer of air is the black layer of asphalt.

There are three light rays coming up from the lower right side. These rays originate from yellow paint on the asphalt, and they all strike the interface at the dashed vertical axis. Ray ***a*** is steeply angled and is intended to represent light that goes straight up. (We just set it slightly to the side of the vertical axis to make it easier to see.) Ray ***a*** essentially goes through the interface between the two layers of air without bending. In fact, Snell's Law predicts that no bending occurs when a ray is perpendicular to the interface, no matter what the temperatures of the air layers are.

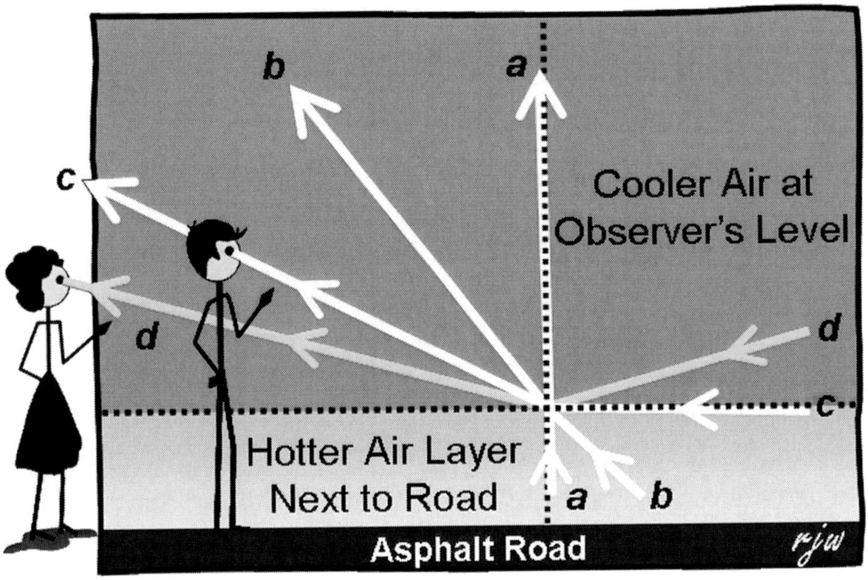

Figure 8–3. Paths taken by light that enters from the right side of a two-layered system of air over a hot asphalt road. Rays ***a***, ***b***, and ***c*** originate from yellow paint on the asphalt and travel up through the interface between the upper and lower layers of air. Ray ***a*** has essentially no bending at the interface. Ray ***b*** bends only slightly. But ray ***c***, that travels nearly parallel to the interface in the hot layer of air, bends upward a whopping 0.36° as it enters the cooler upper air layer. This is an illustration of the extreme refraction that occurs when light travels almost parallel to an interface between layers of air at different temperatures. Ray ***d*** does not pass through the interface. Rather, it is entirely reflected by the change in air temperature at the interface, because its propagation angle is less than the *Minimum Angle.*

Ray ***b*** has a much shallower angle to the interface. As it crosses the interface, ray ***b*** bends slightly upward toward the top layer with the larger IOR. Even though there is some bending, the amount is small. It turns out that for light propagation in air, the beam is not bent very much until the incident angle gets close to being tangent to the interface, say less than two degrees.

To illustrate the increased bending, let's look at ray ***c*** in the lower layer as it moves along the interface from right to left. Snell's Law shows that the bending of ray ***c*** increases substantially as its angle to the interface approaches zero. When the ray from our 140°F road becomes tangent to the interface, it is refracted (bent) by 0.36°.[4] That is the angle of the ray as it enters the denser air layer on the upper left side of Figure 8–3.

This is a phenomenally large bend! At just this single interface, ray ***c's*** light is bent more than 70% as much as the ray we illustrated in Figure 8–1. Remember, that the light in Figure 8-1 traveled downward 20 miles through the atmosphere and covered a 400-mile long path to achieve a bending of approximately half a degree.

Notice that all possible sources of light that might come from the yellow stripes on the road would have an angle between the two extremes of ray ***a*** (which is perpendicular to the interface) and ray ***c*** (which is parallel to the interface). But when the rays emerge from the interface in the upper left quadrant where the two viewers are, they do not cover all possible angles in that quadrant. Ray ***a*** is still perpendicular to the interface, as it was in the hot air layer below. But ray ***c*** is now angled up from the interface by a small amount (0.36°) instead of being parallel to the interface.

---

[4] See *Snell's Law* in the *Glossary* for calculation of the 0.36° angle.

The man's eyes in Figure 8–3 are up at a height where he sees light coming from ray ***c***, which is 0.36° up from tangent to the interface. If he were to walk forward and kept looking at exactly the same spot, he would continue to see light from the yellow stripes. However, the yellow ray he sees would vary with location. As he moves right, his eyes would intercept rays ***c***, ***b***, and ***a***, as well as all intermediate rays not shown in the figure. If he were to back up just a little bit from his current position in Figure 8-3, he would no longer see any yellow light. All the yellow rays would pass over his head.

But when the man backs up and no longer sees yellow stripes, he still sees something. We know that from our own experiences. When we look at a very shallow angle to a surface, we see something! To answer the question of what the man sees, let's consider the woman in Figure 8-3.

She is a little shorter than the man. And she's farther from the spot where the rays enter the upper layer. Her combination of height and distance from the spot produces an angle to the spot on the interface that is less than the 0.36° angle leading up to the man's eyes.

What is seen by her, and anyone whose eyes are within the small angle up from the interface, is ***reflected*** light as illustrated by ray ***d*** in Figure 8–3. Notice, that ray ***d*** doesn't cross the interface. It originates in the cooler upper layer of air and ***reflects*** back up into the cooler air. No energy is transmitted into the lower layer.

The woman and man in Figure 8–3 are both looking at the intersection of the interface and the vertical axis — the spot where all the light rays emerge. But, they see different things.

What the man sees provides us with our first rule for unraveling what happens inside a mirage —

Rule 1: ***When light shines from a less dense air layer toward a denser air layer, there is a Minimum Angle[5] relative to the interface at which the light enters the second layer.***

The *Minimum Angle* is achieved only when the ray in the less dense layer (bottom in our case) travels parallel to the interface. At angles closer to the interface, **NO** light comes from the less dense lower layer.

What the woman sees illustrates the condition described by our second rule —

Rule 2: ***When light shines from a more dense air layer toward a less dense air layer, and the angle of the impinging ray is less than the Minimum Angle, the light is reflected and does not enter the less dense layer.***

To give some feeling for the implications of Rule 2 assume that ray ***d*** has an incident angle of 0.25° up from the interface. In turn, the reflected light departs to the left of the reflection spot at 0.25° up from the interface. The woman could certainly see light from ray ***d*** if she were in the right location. But she'd need to be a long way off to the left. If the woman's eyes were five feet above the interface, she'd need to be 1,146 feet away

[5] See *Snell's Law* in the *Glossary* for the calculation of the value of the *Minimum Angle*.

to see the reflected light ray.[6] That's well outside Figure 8–3.

With these fundamentals of both ***refraction*** and ***reflection*** in hand, we can apply Rules 1 and 2 to explain mirages in general, and the one shown on the book cover in particular.

A mirage can occur when the sun heats a long, flat surface of the earth to a temperature well above the temperature of the overlying air. The earth, in turn, heats the air closest to it. And that air heats the next layer above it, and so forth. These extremely hot layers of air are confined to a few inches above the earth.

If light then shines through the heated air next to the surface at a very shallow angle above the earth, the light is ***reflected*** back upward (just like ray ***d*** in Figure 8–3). If the light in question happens to come from a patch of blue sky way off in the distance (and your eyes are at the same shallow angle to the road as the light from the blue sky), you see what appears to be a pool of blue water. But what you are really seeing is a reflection of the blue sky.

Mirages of the sky are commonplace around Marfa. The long, flat, straight stretches of asphalt road provide ideal conditions for their formation. Automobile headlights show up very well in the heated air, and you will see car headlights reflected at you as you drive in the region.

To see the classic form of mirage that looks like water, the horizon of the road must be free of objects. The mirage conditions are even better if the road is rising somewhat. These conditions give the blue sky the best opportunity to shine down

---

6 Your height and your distance from the point of reflection need to match the 0.25° angle of the light ray up from the interface.

nearly tangent to the road, and then *reflect* up to your eyes. Driving south from Van Horn on US90 toward Marfa, these optimum conditions occur frequently as the terrain rises near the Ryan Flat shown on the map in Figure 2–1.

One of our favorite mirages in the Marfa plain area lies just east of Marfa on US90. It is the mirage shown on the cover of this book. Unlike most highway mirages, this one distorts the image apparently from left to right instead of just reflecting a somewhat distorted, inverted, replica of the sky or an automobile.

The cover photograph was taken from a location on the south side of US90 about 4,960 feet (0.94 miles) west of Nopal Road. This mirage can be seen on just about any clear, dry, warm day. Several features are evident. In the background and to the left, is the readily recognizable Paisano Peak. Nopal Road intersects US90 about half way up the photo, where the truck is seen entering the highway. And, in the foreground, there are two low spots on the road. The mirage seems to be at the bottom of a hill that rises up from the closest low spot. The yellow center stripes on the road appear to swerve across the road at the mirage. And, on the left side of the mirage, the desert seems to encroach on the road.

Two questions come to mind when looking at the mirage on the book cover — Why don't you see features that you *know* are there? And, why do you see features that you know are *not* there?

These two questions have different answers. But both are explainable from the previous discussions about refraction and reflection.

Let's take the first question — Why do the yellow stripes down the center of the road stop coming toward you at the same place where yellow stripes appear to extend across the road from left to right? If you were to move forward from the location where the photograph was taken, the missing yellow stripes would appear where you would expect to see them. So, you know they are really there. Why, then, aren't they in the photograph?

In this situation, the air at the surface of the asphalt is hotter than the air above it. Light from every portion of the yellow stripes comes from warmer air at the asphalt up into cooler air. As we discussed earlier, when light comes upward from warmer air into cooler air, it is *refracted* (bent) *upward* at the interface. Rule 1 above tells us that light traveling upward into cooler air must enter the cooler air at an angle greater than, or equal to, a small *Minimum Angle* — just like light ray ***c*** traveling from right to left in Figure 8–3.

You can only see this light if your height and the distance to the source of light (in this case, a yellow stripe) make an angle greater than the *Minimum Angle*. Otherwise, light from a feature on the road within the mirage region escapes from the heated air at an angle that is above your eyes. So, like the man in Figure 8–3 who couldn't see the yellow rays after he took a step backwards, you can't see the yellow stripes that continue straight down US90 from the position where the photograph was taken. Light from those stripes passes over your head.

Now for the second question, which is the more difficult of the two — Why do we see features like the desert and yellow stripes that aren't where they should be?

The answer comes from our discussion of *reflected* light. In this case, we don't see the sky reflected as a pool of water on the road. But instead, we see the reflection of the nearby features of the terrain. There's nothing magical about seeing nearby terrain instead of the sky in the mirage. It's simply a consequence of the slope of the road at the mirage and the angles to the viewer and land features.

We'll illustrate the process. Figure 8–4 shows an enlargement of a section of the cover photograph, and Figure 8–5 shows two graphs that are accurately scaled renditions of the terrain along US90 at the mirage.

Figure 8–4. Mirage on US90 near Nopal Road. The photograph shows the complicated mirage features.

At the bottom of Figure 8–5, Graph A shows nearly a mile of US90 — almost as much as in the cover photograph. Zero on the horizontal axis is the location where the photograph was taken. That position is designated point ***g*** in both Figure 8–4 and Figure 8–5. Nopal Road intersects US90 just off the right edge of Graph A. The vertical axis in Graph A is the elevation and runs from zero to 30 feet above a reference value of 4,780 feet asl. The two low spots in the road are clearly evident. They are shallow depressions of only a few feet in the 4,000-plus feet of roadway.

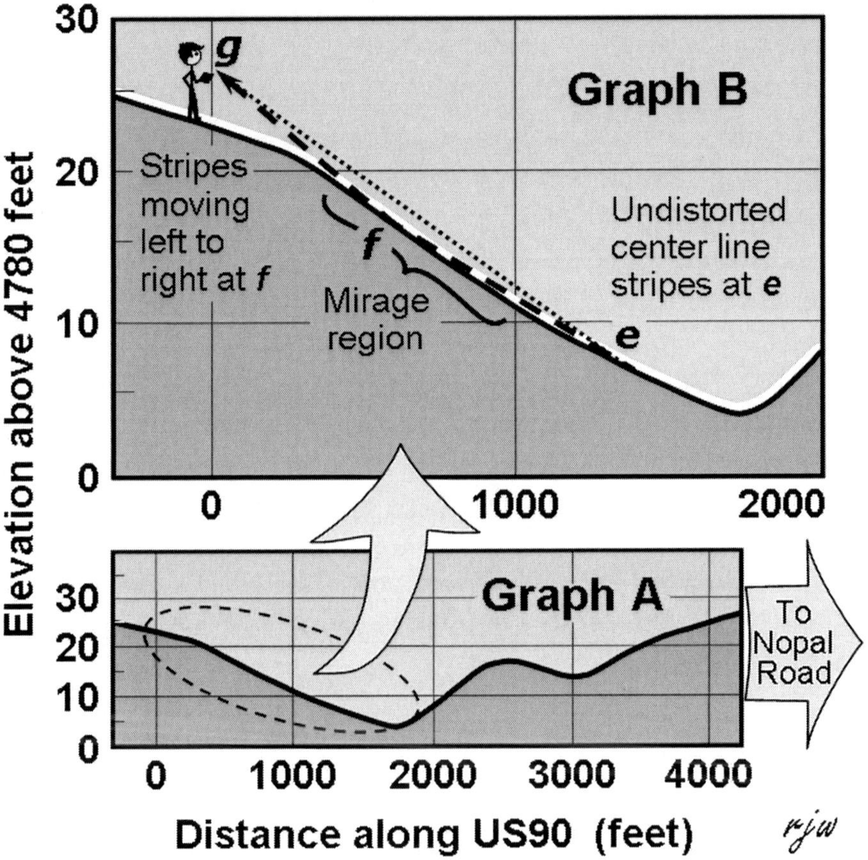

Figure 8–5. *Graph A at bottom* shows the elevation along US90. *Graph B* shows the detail of the elevation in the mirage region. The white band along the road is meant to symbolically represent the very hot air adjacent to the asphalt road. However, it is not to scale like the road elevation is. In reality, the white band is only about half an inch thick, and thus much too thin to illustrate accurately.

The specific region of the mirage (dashed oval) is shown in greater detail in Graph B at the top of Figure 8–5. Here, we have enlarged the vertical scale, placed a person at ***g*** where the camera was, and added a hot-air layer in white above the road. (Note that the white layer is not to scale.)

Let's look at the area in the photograph in Figure 8–4 where the two yellow center stripes seem to run from left to right. We've drawn a dashed white line connecting the viewer's line-of-sight from ***g*** to both the left-to-right deflected yellow stripes and the undistorted yellow stripes beyond. The two places where this dashed line crosses the yellow stripes have been labeled ***e*** and ***f***. These two locations are also shown in the elevation plot in Graph B of Figure 8–5. Location ***e*** is beyond the mirage region, while ***f*** is within the mirage region.

Incredible as it may seem, location ***e*** is what you are actually seeing when you perceive something at location ***f***. In this mirage, you are seeing the road stripes at location ***e*** twice!

First, you see them in the expected location as light travels essentially directly from ***e*** to your eyes at location ***g*** along the upper dotted line in Graph B. In this case, the light is traveling through air ***above*** the highly heated portion (shown in red) that is closer to the road.

But there is another ray of light that follows the dashed line ***e*** to ***f*** to ***g*** in Graph B. This ray is analogous to ray ***d*** in Figure 8–3 that had an incident angle less than the *Minimum Angle.* It travels through cooler air a few inches above the road until it hits the really hot air layer at the asphalt ***(f)*** and *reflects upward.*

It then continues on to your eyes. The ray behaves as specified by Rule 2 and reflects because it travels from ***e*** to ***f*** at a very shallow angle that is less than the *Minimum Angle.* This reflection of the stripes that are really at ***e*** is perceived at a lower angle than the direct image of those stripes. So you see it at a lower position in the photograph in Figure 8–4.

This general description of how features, that are remote from the mirage, show up in it can also be applied to explain how the desert on the left side of the photo appears to spread across the road. Look closely at the desert mirage in the photo. With some difficulty, you can identify specific objects from the left side of the road. The shiny guard rail is evident in the near edge of the mirage. It is in a straight line between the guard rail in the distance and the camera's location. The explanation for the two images of the guard rail is identical to the one for the yellow center stripes on US90.

The mirage photograph on the cover is an excellent example of what occurs when light is *refracted* (bent) and/or *reflected*. But these phenomena are not confined to daytime. Since the Marfa lights are nighttime phenomena, we need to examine how conditions change after dark. We'll turn to that subject in the next chapter, *Larger Than Life*.

—§—

8 — Sources

[1] George Peacock, ed., *Miscellaneous Works of the Late Thomas Young, Volume II* (London: John Murray, 1855), 53.

[2] James Hodgson, *A System of the Mathematics, Printed for Thomas Page* (London: William and Fisher Mount, 1723).

[3] Nathaniel Bowditch, *The New American Practical Navigator, 5th Edition*, (New York: Edmond M. Blunt, 1821). [First edition published in 1802.]

Nathaniel Bowditch, *The New American Practical Navigator, 16th*

Edition, (New York: E. & G. W. Blunt, 1846).

[4] Henry Adams, *Practical Surveying and Elementary Geodesy* (New York: The MacMillan Company, 1913), 124.

[5] James Bunnell, *Night Orbs* (Cedar Creek, TX: Lacey Publishing Co., 2003), 37-38.

[6] Jack A. Stone and Jay H. Zimmerman, "Index of Refraction of Air," in Engineering Metrology Toolbox, Physical Measurement Laboratory, NIST, http://emtoolbox.nist.gov/Wavelength/Documentation.asp

Goat Mountain

9 — Larger Than Life

And a little while after night came, a full moon rose up over the mountains and it was still a beautiful sight to see those mountains by moonlight. They looked like they had just moved up a little closer.

– From memoirs of Robert Reed Ellison[1]

WHAT MADE THE MOUNTAINS Robert Ellison saw look like they'd moved closer? In order to answer this question, we need to understand nocturnal light propagation. Specifically, what makes light propagation different at night from what occurs during the day?

The answer is at once simple and complex — air temperature.

Remember from the previous chapter that refraction of light by the air is controlled by a parameter of the air called the *index of refraction*. In turn, the index of refraction is determined by five factors — wavelength of the light, air pressure, air temperature, relative humidity, and carbon dioxide concentration. Of the four environmental factors, the one that has the greatest influence on the index of refraction is the air temperature. As a consequence, changes in the air temperature along the path of a ray of light have the greatest influence on the refraction of the light ray.

While air temperature is the most powerful factor in determining refraction of light by the air, it is also the most difficult to determine because it is tied up in a very complicated phenomenon called the *nocturnal temperature inversion.*

If we are to understand nighttime light propagation in mountains, valleys, and plains between the valleys in the greater Marfa region, we need to understand the nocturnal temperature inversion. That understanding will enable us to explain why the mountains, that Robert Ellison saw when he camped on the Marfa plain in March 1883, appeared to have moved closer.

Everyone knows that after sunset the air begins to get cooler. But, the air does not cool spontaneously. Air cools because it rubs up against something that is cooler than it is — the earth. Yes, the earth is what drives the cooling in the evening.

During the day, the sun showers the earth with radiation in the visible range from blue to red, as well as in many other wavelengths of light that we can't see. This radiation warms the earth's surface. In the summer, it warms the surface a lot, as you can readily confirm if you try walking on a concrete or asphalt road in your bare feet. But after the sun sets, the radiation on to the earth's surface stops; and the earth begins to cool rapidly.

Even on a windless day, air is not motionless; it is generally churning about. This churning helps cool the air at night. Imagine a warm parcel of air that comes in contact with the earth. As it rubs up against the earth, the air parcel gives up some of its energy to the cooler ground. After that contact, the air is cooler, and the earth has gained a little bit more energy.

The parcel of air which was cooled by contact with the earth continues to churn about bumping into other bits of air and transferring energy. Hotter air that the parcel encounters will transfer energy to the initial parcel. Cooler air receives energy from the initial parcel. This exchange process goes on everywhere close to the un-illuminated earth. The result is a steady transfer of energy downward, warming the earth's surface and a cooling of the air upward from the earth. The heat transferred downward to the earth is subsequently radiated up into space as the earth continues its overall cooling.

If the winds after sunset are sufficiently strong, the cooled air will be swept up and distributed to high levels of the local atmosphere — mixing the warm parts with the cool parts. But if the winds are weak — say 5 mph or less — the air cooled by the earth tends to hang around closer to the surface of the earth. Since the cooled air is denser than warm air, it has a natural tendency to sink below warmer air. If there were no winds at all, the cool air would be even more inclined to stay below the warm air and near the surface of the earth. If this cooling process continues for a few hours in low wind conditions, a substantial pool of cold air develops near the ground.

"Near the ground" can mean many things depending on the entirety of conditions where the cooling occurs. Cooling is influenced by winds, humidity, vegetation, time since the last rain (moisture content of soil), elevation, proximity to bodies of water, and cloud cover. For the Marfa region, the conditions are semi-arid, low humidity, low vegetation, low soil moisture content (no recent rain), 5,000 feet elevation (lower air density), no streams or near-by lakes, generally clear skies, and winds that die down after sunset.

Under these conditions, it's reasonable to imagine that the first 300 feet of air above the ground is cooled below the temperature at the 300-foot level. The amount of cooling depends on the time of year. The span of air temperatures from the ground up to where the air stops warming is called the *nocturnal temperature inversion* (NTI).

Many people have measured the NTI over the last 50 years.[2] Their efforts have illustrated how it is dependent on the parameters we've just listed. One of the best treatments of the NTI and how to characterize it for mathematical modeling comes from the Ph.D. dissertation of a mechanical engineer, Johannes Kloppers.[3] Fortunately for us, he used air-temperature data collected above a relatively arid, sparsely vegetated plain around Lephalale, South Africa.

The terrain there is much like Marfa, being 3,000 feet above sea level (asl) with surrounding hills and mountains extending some 400 to 1,000 feet higher. Marfa and Lephalale are at comparable distances from the equator. Marfa is at 30° north latitude, and Lephalale is at 24° south latitude. So they experience comparable amounts of solar radiation, but their annual temperature cycles are about six months out of phase with one another.

Kloppers wasn't interested in the NTI for its own sake. He simply wanted to know how to model the air so he could develop design equations for cooling towers which employ atmospheric air forced through them. His air-temperature data were collected from a 96-meter (315-foot) tall tower. Nighttime air temperature measurements were taken at eight places along the tower every six minutes between March 1994 and the end of January 1995. The data corresponding to NTIs were culled out for analysis.

Kloppers found that the temperature difference between the top of the inversion and the air temperature one meter above the ground could be as large as 10° to 12°C during the South African winter. During the summer, the difference in temperatures might be only 6°C. That's equivalent to a difference of 18° to 22°F in winter and 11°F in summer.

A specific example of an inversion measured at Lephalale is shown in Figure 9–1. These measurements were made on the night of 8 June 1994 (winter time in Lephalale). The smooth curve through the data points on Figure 9–1 is from an equation the authors created to give a good fit to Kloppers' measurements.

Notice how the temperature rises as the height of the measurement instrument on the tower increases from near ground level to 315 feet above the surface. Clearly, the colder air stays pooled near the ground.

Given the similarities between the Lephalale and the highlands of the Big Bend, the NTIs in the greater Marfa area should follow the same cyclical variations throughout the year. Therefore, we would expect the biggest inversions on the Marfa plain in the November–February winter period, and the smallest inversions during the summer period from April to September.

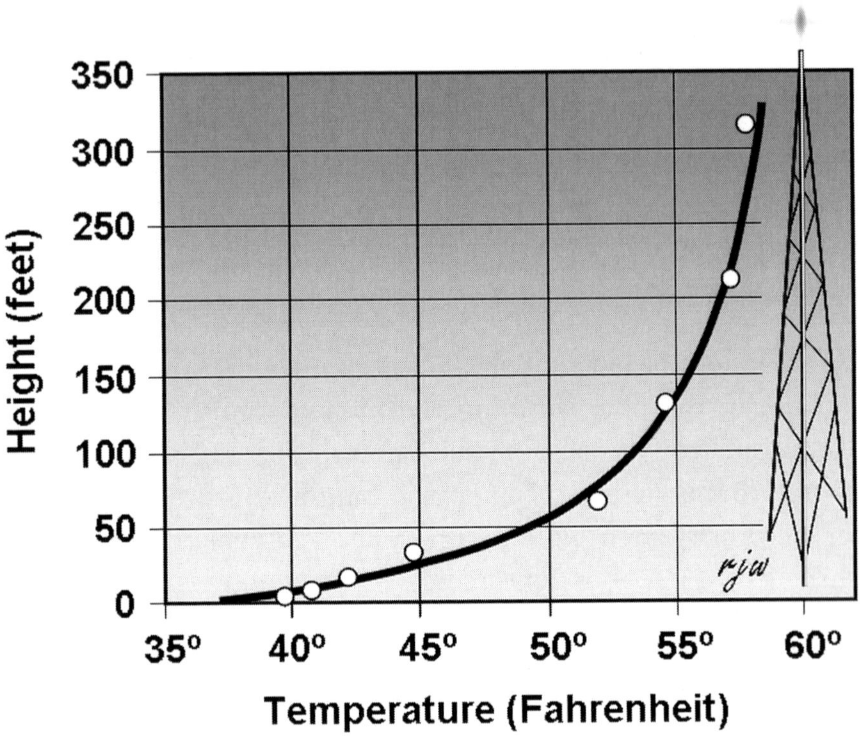

Figure 9–1. Air temperature versus height for the first 315 feet above the earth at Lephalale, South Africa on the night of 8 June 1994. The dots are measured values at eight locations along the measurement tower. The smooth curve is the authors' mathematical fit to the data.

But how can we see if this is true? How can we enhance our confidence that the South African NTI data is applicable to the Marfa plain?

It would be wonderful if we had air-temperature data from a 300-foot tower on the Marfa plain. Unfortunately, such data does not exist. Undaunted, we were determined to examine the correlation of the Marfa-plain data with that of Kloppers. To do so, we had to come up with a portable tower and equipment to record the temperature.

On 18 November 2007, and again on 28 December 2008, the authors erected a 20-foot mast at the Marfa Lights Viewing Center. We attached temperature sensors/loggers at the top (20 feet) and at the one-foot level. On both occasions, data were taken for several hours following sunset.

It got cold in a hurry! In about an hour, the November measurements showed an air temperature difference of about 9°F between the **one-foot** and **20-foot** levels. The December temperature difference was about 11°F.

These two measurements from the Marfa plain can be contrasted to the air temperature difference between the one-foot and 20-foot levels above the earth of the South African data. Kloppers' data shows a difference of about 7°F. In the first 20 feet, the Marfa plain NTI achieved an even higher differential of temperatures than the South African data! This is a very favorable comparison of data and makes a strong case for our using the solid curve in Figure 9–1 in our calculations and analyses of light propagation after dark.[1]

Because cold air collects in low areas, the terrain influences how and where the colder air pools on the earth's surface. So, before we can apply our knowledge of the NTI, we need to define a model for the surface of the Marfa plain. That model, in combination with the NTIs that develop on the Marfa plain, will permit us to analyze how light is refracted as it

---

[1] Later in this book, we'll find cases where the magnitude of the NTI is not the same as that shown in Figure 9-1. In those situations, we'll lessen the magnitude, but preserve the shape of the NTI depicted by the solid curve in Figure 9–1.

propagates through the varying air temperature.

As its name suggests, the Marfa plain is very flat. From the general vicinity of Antelope Springs, you can travel southwest on a ***relatively flat, downward slope*** about six and a half miles toward Chinati Peak before encountering a major break in the plain. The break occurs just east of the Alamito Creek and the old Kansas City, Missouri, and Orient (KCMO) Railroad.

Make the same excursion southward, and your trip would be ***relatively flat and level*** for about seven miles before you encounter the hills south of the plain. Head toward Goat Mountain, and it would be ***relatively flat and upslope*** for about ten miles before the mountain rises. Use Paisano Peak as your destination, and the trip would be ***relatively flat and upslope*** for about eight miles until foothills arise.

These observations would certainly make you want to say that the Marfa plain is pretty much flat and level. You'd be right. Overall, the Marfa plain slopes downward from about 5,000 feet asl in the northeast corner to about 4,500 feet asl at the southwest corner. This decline occurs over a distance of about 15 miles, yielding an average slope of less than half a degree. That slope is so slight that we don't need to worry about any tendency for the cold air generated next to the earth to slide down slope.

The other factor of the terrain, its flatness, is of greater concern. As we've shown above, the NTIs that will develop on the Marfa plain are expected to have vertical dimensions in the hundreds of feet — say 300 feet. If the terrain on the Marfa plain had little hillocks that popped up regularly, rose up a few hundred feet, and then went back down, they would essentially poke holes in the layer of air that forms the NTI. That would be a problem because we need long, uninterrupted distances of light propagation in an NTI to refract the light by an appreciable amount.

Once again we are in luck. The Marfa plain goes on for miles and miles with little change in flatness. There are occasional hillocks, but they are few in number. If hillocks are not a concern, how about a slow undulation of 300 feet from peak-to-trough? Again we're in luck. Out in the main portion of the plain, the general undulations of the terrain are on the order of 50 feet peak-to-trough, and occur over distances of miles.

This means that the NTI will tend to cover the terrain like a blanket and not just collect in isolated low points. From these facts about the surface features of the Marfa plain, it is safe for us to adopt a flat-earth model for the plain and its overlying NTI.

Now, let's see how the terrain and the NTI work together to affect the refraction of light on the Marfa plain at night.

In our first illustration, we'll consider a general hill-mountain structure 13 miles distant from an observer who is located in the vicinity of Antelope Springs. We'll examine light locations on the hill up through the first 700 feet. These heights and distances don't match any specific location on the Marfa plain, but they are close to what you would see when looking in the direction of the hills around the western entrance to Rustlers' Gap. (See Figure 2–1.) Because there are so many lights produced in this region, we were motivated to choose an approximation to that view. Also, the highest peak visible in that direction is about 900 feet higher than Antelope Springs.

Figure 9–2 illustrates the scene we've chosen for our

model. The observer on the lower left of the graph is standing at Antelope Springs. That is, Antelope Springs is at the origin of coordinates. Our observer is looking 13 miles across the plain to the "cartoon" hill-mountain rising up 800 feet on the right side of the graph. We've exaggerated the shape of the mountain for purposes of illustration.

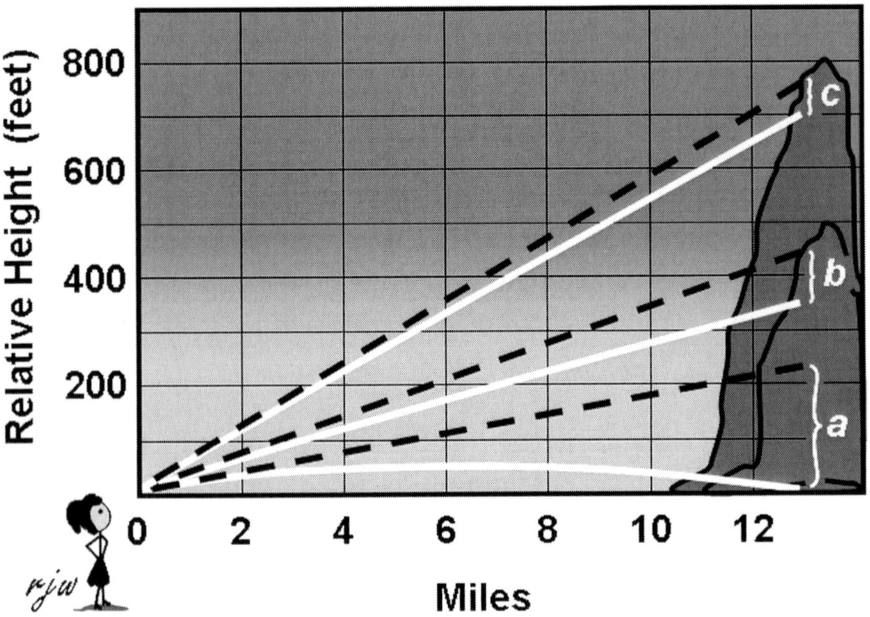

Figure 9–2. Illustration of light rays propagating across the Marfa plain when there is a strong nocturnal temperature inversion (NTI) present. In the case shown, the NTI extends from the ground upward about 400 feet into the air. Solid white lines show the paths followed by light rays. Dashed black lines show the apparent directions to the sources of the rays. Strong refraction of the bottom two light rays (***a*** and ***b***) is evident by the curvature of their paths.

The lighter shading on the lower portion of the graph represents the colder temperatures of the air pooling nearer the earth's surface. The cooling selected for this analysis is that of the NTI shown in Figure 9–1 and extends approximately

300-400 feet above the plain.

The three white light rays (***a***, ***b***, and ***c***) emitted by the terrain of the mountain are shining toward our observer at Antelope Springs. Although the terrain itself doesn't glow, it does reflect light from other sources. The lights could be from a full moon and thus are very bright. Or, they could be from reflected starlight and be rather dim. It doesn't matter where the light originates initially, nor that the light reflects in all directions — upward, downward, to the sides, as well as toward our observer. The only light that is visible to our observer is the light that is reflected toward her eyes.

We've colored the light rays white just to highlight them. In fact, they have the same color as the mountainside. Each dot of mountainside projects its own colored image off into space. The paths followed by the lights that reach our observer are the ones of interest to us.

Light rays ***b*** and ***c*** arc downward from their sources on the mountainside to the observer. Initially, light ray ***a***, the lowest of the rays, shines upward before curving over into its downward arc. To calculate the shape of these curves, we used Snell's Law, (see *Glossary*) along with the actual, numerical values of the NTI at every segment and height along each light ray's path. To determine how the angle of the light ray changed as it passed from one segment to the next, we divided the light's path into 1,000 short segments (connected end-to-end) and solved Snell's Law at every interface between two segments.

For the observer to see light from a source on the mountainside, the angle of the light when leaving its location must be just right for the resulting arc of light to return to earth at the observer's location. In our calculations, we varied the initial launch angle of the light into the first segment of its path and let the mathematics select the rest of the path. When the launch angle was just right, the light reached the observer correctly.

As Figure 9–2 reveals, the three white light rays arrive at the observer with different final angles of travel. That final angle is the only one the viewer can discern, and the viewer thinks the light she sees came from the direction of that final angle. We used this final angle to project back toward the mountain to where the observer perceives the light to have originated. These projections are illustrated by the dashed, ***straight***, black lines.

In every case, the light of the mountainside appears to have come from a location higher up the mountain than the actual height of the light source. When light appears to come from higher than the actual source location, the light is said to be *looming*.

The apparent increases in height vary with location. Location ***a***, at the base of the mountain, has a whopping 221 feet increase in apparent height from its true location! That's 25% of the height of the mountain. Locations ***b*** and ***c*** have increases of 92 and 53 feet, respectively.

All the looming increases arise because the light traveled, to some extent, through the NTI and was refracted as it propagated. Path ***a*** was totally within the NTI and had the full length of 13 miles to be repetitively bent as it passed through the various layers of air with their varying temperatures. So, ray ***a*** got the largest apparent increase in height. Paths ***b*** and ***c***, each had a portion of their propagation that was above the NTI and a portion that was within the NTI. Path ***c*** had less distance within the NTI than ***b***, and ***b*** had less than ***a***. Accordingly, their apparent height increases followed in the same relation. The more time the light spent in the NTI, the greater the looming effect.

Now, imagine that we're not just seeing a few reflected light rays. The light being emitted comes from each and every point on the side of the mountain. In fact, the perception of the observer from all those points of light is the image of the mountainside. Each of those points of light arcs down to the observer just like the three points of light in Figure 9-2. And — *voilà* — the mountain appears larger and nearer to us than we would expect. This is the looming phenomenon, and it's not a new phenomenon. Consider the story below.

On July 26, 1797, William Latham, Fellow of the Royal Society, witnessed the looming phenomenon[2] over the English Channel.[4] From his location in Hastings, England, he could see the White Cliffs of Dover on the English side of the English Channel and numerous towns on the other side of the channel in France. Dover is 40 miles away from Hastings, and the towns in France were from 50 to 60 miles away. These sightings were made by Latham and many others in Hastings that day with their unaided eyes. In Latham's correspondence to the Royal Society, he wrote —

> *I pursued my walk along the shore to the eastward, close to the water's edge, conversing with the sailors and fishermen upon the subject. They, at first, could not be persuaded of the reality of the appearance; but they soon became so thoroughly convinced, by the*

2 The cold water of the English Channel combined with the unusually warm air above it to produce a temperature inversion in the air.

> ***cliffs gradually appearing more elevated, and approaching nearer,*** *as it were, that they pointed out, and named to me, the different places they had been accustomed to visit. . . .*[4]

We've added emphasis to the words about cliffs appearing more elevated and approaching nearer. It was well known to the scientists[3] of the time that looming made objects appear to be closer than they really were. People perceive objects to be closer because they have a natural gauge for how near something is to them based on how large it seems to be. You can find this phenomenon explained in many sources produced from the time of William Latham's sighting up to the present.

Just as looming was a real phenomenon on the English shores in 1797, it is equally possible that looming existed on the Marfa plain on the night in March, 1883 when Robert Reed Ellison camped there. Ellison was heading for his father's range on Alamito Creek near Plata (see Figure 2–1). But, on the day he drove his cattle out of Paisano Pass, his immediate destination was Antelope Springs. In his memoirs, he wrote about what he saw during his first night on the Marfa plain —

> *And a little while after night came, a full moon rose up over the mountains and it was still a beautiful sight to see those mountains by moonlight. They looked like they had just moved up a little closer.*[1]

We have seen above that a reasonable NTI can make mountain sides 13 miles away appear to be more than 200 feet higher than reality. Robert Ellison's description of the moun-

---

[3] William Latham was indeed a scientist in his time. He was a Fellow of the Royal Society of London. The Society was dedicated to research and discussion of the natural sciences.

tains moving closer suggests that the Marfa plain experienced an NTI.

This story could end with the observation that an NTI might have occurred on the night Ellison camped on the Marfa plain. But, that would still leave the explanation as more conjecture than science. It would also leave the story in the realm of mythology. More information would be needed to clarify what happened that night.

To discover if an inversion occurred in March, 1883, we needed to establish the date and meteorological conditions on the night Ellison camped near Antelope Springs. That would not be easy. Ellison's memoirs place the date several weeks into March and on a night with a full moon that came over the mountains "a little while after night came". He makes no mention of the weather or how cold it was that night.

There was a full moon on March 23rd. For a few days before and after that date, the moon would have appeared large enough for someone to call it *full.* We can eliminate the dates before the 23rd because on those days the moon rose before the sun set, making it impossible for Ellison to view a moonrise after dark. But from the 23rd through the 26th of March 1883, moonrises occurred anywhere from 14 minutes after sunset to 2 hours and 49 minutes after sunset. These dates are candidates.

Sunset for this period in late March was between 7:08 p.m.[4] and 7:10 p.m. It would take approximately an hour for the moon to get high enough in the sky to be seen above the mountains. That means the moon would have come into Elli-

[4] Unless otherwise noted, all times are central standard time (CST).

son's view over the mountain at approximately 8:22 p.m. on the 23rd, 9:14 p.m. on the 24th, 10:07 p.m. on the 25th, and 10:59 p.m. on the 26th. While the last time may appear a little late to fit Ellison's description, there is really no way to know what he meant by "a little while after night came".

The next question was whether the meteorological conditions were right on any of these dates for an NTI to develop. In 1883, the United States War Department was responsible for creating a national weather map at 7 a.m. EST each day. These maps have been digitized and are accessible from NOAA Central Library Data Imaging.

The scientific sophistication of the east coast of the United States had not reached the frontier of Texas in 1883. But they did have equipment in Fort Stockton with which they took daily readings of temperature and wind speed and direction at 6 a.m. On the 24th to 27th of March 1883, they reported temperatures of 44°, 48°, 46°, and 48°F, respectively. We know from historical data that, three hours after sunset in March, the temperature on the Marfa plain can be as much as 15° colder than at 6 a.m. in Fort Stockton the next day. So, it could have been very, very cold at Antelope Springs. Certainly, it could have been cold enough for a strong NTI to develop.

While it was possible for a strong NTI to occur, was it likely?

Four meteorological conditions are needed for the formation of a strong NTI — no recent rain, low wind, clear sky, and low humidity. Fortunately, the War Department's weather maps described these conditions. They showed that it hadn't rained since March 16th. So that condition (dry soil) checks out. There were low winds and clear skies for March 24 to 27 —

two more requirements met. Clear skies can be interpreted to mean "low enough" humidity. So the basic requirements for a substantial NTI to develop were in place.

Finally, the War Department's weather maps for the 25th and 26th were especially intriguing.

On the 25th, the national weather map showed a large low pressure system over central coastal Texas and warm temperatures through the Midwest. The following morning brought an entirely different picture. The 7 a.m. EST map for the 26th showed a raging cold front dropping down the central plains of the country. Temperatures at the Canadian border were below zero, and the temperature at the Red River just east of the Texas panhandle was 38°F. All the winds behind the front were straight out of the north.

The front hadn't reached Fort Stockton on the morning of the 26th. The skies were clear; the winds were light and from the east with a 6 a.m. temperature of 46°F. One hundred and fifty miles further east, Fort Concho showed 43°F and cloudy, with 8 mph winds out of the northeast.

As this weather event moved further south during the day, it is likely to have produced a very cold night on the Marfa plain on March 26, 1883. Even if the direct impact of the front didn't reach the plain itself, the front probably caused cold winds from the Davis Mountains to blow southward earlier in the evening than normal. Those winds can bring very cold temperatures down from the heights of the peaks as we will see in *Chapter 11 – Aragon ~ Lights Below the Chinatis.*

Adding all of this together, we conclude that it is not only plausible, but highly likely, that a substantial NTI occurred on the night of Robert Reed Ellison's trip across the Marfa plain late in March 1883. And, Ellison, like William Latham in 1797, was witness to the phenomenon of *looming*.

Robert Ellison's story is only one example of the effects of an NTI on our perception of light. In the next two chapters, we will continue our examination of refraction and the NTI's role in the mystery that *is* the Marfa lights. This will lead us to investigate the most elusive of Marfa lights mysteries, and answer the question — Why are lights seen to move upward sometimes?

—§—

9 — Sources

[1] Cecilia Thompson, *History of Marfa and Presidio County, Texas 1535-1946, Volume 1, 1535-1900* (Marfa, TX: The Presidio County Historical Commission, 1985), 194.

[2] Heinz Lettau and Ben Davidson, (eds.), *Exploring the Atmosphere's First Mile* (London: Pergamon Press, 1957).

[3] Johannes Christiaan Kloppers, "A Critical Evaluation and Refinement of the Performance Prediction of Wet-Cooling Towers." Ph.D. dissertation, University of Stellenbosch, Stellenbosch, South Africa, December 2003.

[4] William Latham, "Account of a Singular Instance of Atmospheric Refraction," *Philosophical Transactions of the Royal Society of London*, 88, (1798): 357-360.

10 — Lights on the One O One Flat

JUST AS DANCING BROUGHT RUTH BOWNDS to a setting where she observed mysterious lights east of Marfa, an interest in geology brought Robert Black and his friend to the southeastern side of the Marfa plain where they witnessed mysterious lights early one morning in December. According to the story recounted by Judith Brueske in her book, *The Marfa Lights,*[1] the adventure began as a field trip.

Black, a geology student at Sul Ross State University, wanted to compare the geology on Goat Mountain to that of nearby Cathedral Mountain. So late Saturday morning, November 30, 1985, he and his friend parked their truck at the base of Goat Mountain and hiked up to the top. At the end of the day, they misjudged how long it would take to descend Goat Mountain and arrived at its western base after sunset. Unable to locate their truck in the dark, the hikers were forced to light fires and spend the night out on the Marfa plain. To occupy their time, they decided to look for mysterious lights. Early the following morning, their quest was rewarded.

The details Black recounted are key to unraveling this mystery. So let's recap them.

Where were they? According to Brueske's account, they were on the west side of Goat Mountain. Black said, "Anyone who knows the Marfa flats, knows that it is flat, featureless, and boring — no geological marker out in the sea of desert, really no way to find your way around, especially in the dark."

Where did they look? They thought their viewing direction was north and northwest. Brueske described the direction as "more or less northwards". The lights appeared several feet above the tops of the creosote bushes where the two were camped.

When were the sightings? The early morning hours of December 1, 1985. According to Brueske's account, the men estimated that their first sighting occurred not long after midnight. They saw the lights about four or five times in total at about one-hour intervals. The lights were only visible for a few seconds at a time. Since the men had no watches, Brueske cautions that "all time estimates are vague".

What did they see? Black described the initial sighting as "a horizontal length of light that had a sort of dancing, vibration movement". The lights appeared like

> *little beams of light that would dance up and down in kind of a wave formation, move across, then they'd*

> *seem to jump straight up vertically, come back down, then dance horizontally, then disappear. ... We were happy to see them, especially that close.*

Black indicated to Brueske that the lights always reappeared in the same location and that the color of the lights ranged from yellows and oranges through red.

What were the weather conditions? Brueske reported that the weather on the day of the hike was warm, the sky was clear, and there hadn't been any rain for a long time. The following morning, there was frost on the ground.

Intrigued by Robert Black's detailed and credible account of their night at the base of Goat Mountain, the authors felt there must be a logical explanation for what the men had seen. The obvious choice was headlights. But headlights coming from where? Could the headlights have been from cars on US90?[1]

Viewing Goat Mountain and the other features of the Marfa plain in Figure 2–1 (*Chapter 2 – The Marfa Plain*), you can tell that a campsite on the western side of Goat Mountain is not close to US90 or any other major roads or population centers. The distance to Marfa is about 22 miles, to the Marfa Lights Viewing Center (MLVC) is about 14 miles, and to Paisano Peak is about 13 miles.

To a traveler headed east from Marfa toward the MLVC, Goat Mountain dominates the southeastern horizon. Where US90 curves just east of Nopal Road and west of the MLVC (see Figure 2–1), both the highway and the Southern Pacific Rail-

---

1 US90 and US67 share the same roadway from Marfa to Alpine. For simplicity, we will refer to this stretch of highway as US90.

road tracks that run parallel to it, are aimed within about 13° of the mountain's western side, which is around 14 miles away.

We know from the *Chapter 6 – "Classic" Marfa Lights* that viewers at the MLVC can easily see the classic Marfa lights which are 14–25 miles in the distance. So, eastbound cars at the curve on US90 (or trains on the tracks alongside) would be within the visible range of the campers. Also, geometry and light intensity would enable headlights to be seen from the base of Goat Mountain.

At first, we thought we had our answer. Unfortunately, other factors ruled these headlights out. The curve is only 0.8 of a mile long. That means that the headlights from a car traveling at 60 mph would be visible for only 48 seconds, and they would move smoothly over the road surface.

This does not match Black's description of lights dancing across, then seeming to shoot up vertically, then dancing horizontally, then disappearing. Also, lights on US90 would have occurred all night long,[2] not just after midnight as Black described. And, Black himself, dismissed the possibility of train lights because the men hadn't heard any trains.

We have to conclude that the campers did not see lights from cars on the curve of US90 — because the local terrain and creosote bushes blocked any view of cars at the curve. Lights anywhere else on US90 would not have been visible to the campers because the highway was either at too much of an angle to the campers, or because of high terrain somewhere along the path from the highway to the campsite.

---

2 The Texas Department of Transportation reported that in 1985, 1,200 vehicles per day (24 hours) made the trip over the section of US90 connecting Marfa and Alpine.[2]

If the lights didn't come from US90, then where did they come from? To answer that, we need to locate where the men were camped.

The map in Figure 10–1 is an enlargement of a section on the right side of the map in Figure 2–1. The upper left region of the enlargement is located just below, and to the east, of Paisano Pass. The road at the top left of the map in Figure 10–1 intersects US90 about four miles north of the point designated ***a***. This junction is just south-southwest of the entrance to the Paisano Baptist Encampment.

At the bottom of the map is Goat Mountain with a summit of 6,608 feet above sea level (asl). The top mesa of Goat Mountain is delineated by a 6,000-foot contour. The mountain descends rapidly from the mesa to its base. Roughly one-half mile to the west from the mesa top, the mountain side is down to 5,400 feet. A half mile further west, the mountain base is down to 5,200 feet. Even after the climbers descended from the cliff faces and the two major contours of the base, they would still be on substantial slope until they reached the road running along the western side of the mountain, more than 1,000 feet below the mesa.

The statement that the hiker's campsite was "on the west side of Goat Mountain" provides only general guidance. The west side of Goat Mountain covers a lot of territory. As illustrated in Figure 10–1, the distance from the southern end of the mountain base to its northern end is about two miles. There is one other problem presented by the statement that they were on the west side of Goat Mountain — the "broadside" direction of Goat Mountain is west-northwest, not west. So, if Black was wrong about which direction was west, he would also be wrong in his description of the direction they looked.

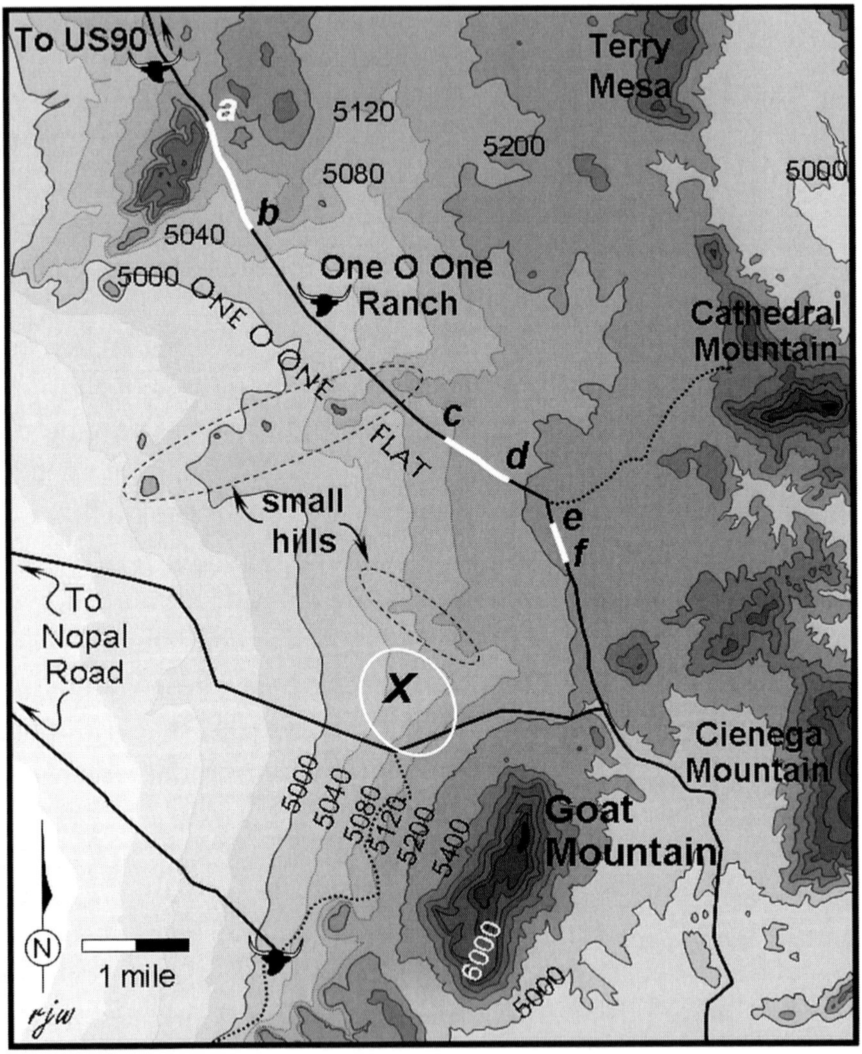

Figure 10–1. Features on the Marfa plain near Goat Mountain. The white oval with the ***X*** at the base of the mountain is the location of the potential campsite of Robert Black and his friend.

The white oval at the northwestern base of the mountain indicates the general area where the two hikers might have camped. This region — which we will refer to as the *camp area* — was chosen for several reasons.

First, it is directly accessible from Nopal Road.[3] Second, the southeast border of the camp area is about as close as someone can drive to the west side of Goat Mountain without encountering the steeper slope of the base. It also encompasses a length of the road along the mountain where the hikers might have parked their truck. Third, it is near the northwest end of Goat Mountain, which appears to be an easier climb than regions farther south along the western side. Fourth, it is about the distance away from the mountain where creosote bushes begin to appear. And finally, proceeding from the southeast side of the camp area to the northwest side, the terrain is relatively flat, as the hikers indicated. Ultimately, the northwest boundary of the camp area was chosen by considering that, while no one would park that far from the base of the mountain, they might inadvertently miss their truck in the darkness and walk that far northwest before giving up the search that night.

With no way to know for certain where the two men camped, we selected a potential location — marked by an ***X*** — for the analyses discussed below. The site is more or less in an arroyo, occasionally fed by rainwater run-off from the east side of Goat Mountain and from the range of hills running between Goat Mountain and Cathedral Mountain. So there should have been plenty of creosote bushes within gathering distance.

The answer to the second question — *Where did they look?* — gives us a significant clue to what occurred the night of November 30, 1985. Black said they looked north and north-

---

[3] Judith Brueske indicated that Black drove out a county road east of Marfa. It is likely to have been Nopal Road.

west. The answer provided by Brueske was "more or less northwards". But in the dark, how did the campers know which way was northerly?

One obvious answer is that as the sun set at 5:55 p.m. CST on November 30th, they observed its lingering glow on the western horizon.[4] Knowing the approximate direction of west, it's easy to determine approximately where north is — just point your outstretched left arm to the west and you are looking northward.

But once the sun had set, it would have gotten dark very quickly because the moon did not rise until 8:29 p.m. CST. For the following two and one-half hours after sunset, the stars were the only illumination. As anyone who has been to the MLVC on a clear night will attest, the stars are brilliantly displayed, and the North Star is readily visible. If Black and his friend knew that the North Star was the outermost one in the handle of the little dipper, it would have been easy to locate the star to get their bearings. Finally, sitting at a campfire, the two men would have seen the moon rise from the east at 8:29 p.m. and move steadily overhead.

With several ways for the campers to orient themselves, we feel confident that they could generally determine north from south. But Black's statement about being west of Goat Mountain is troubling. And now that we've selected a location for their campsite, there is yet another possibility of error in determining where they looked.

If they thought that east was the upslope direction from their campsite to the heights of Goat Mountain, then what di-

---

[4] On November 30, 1985, the sun set 15° south of true west.

rection would they think is north? Look at Figure 10–1 for the answer. The direction you'd call *north* would actually be about northeast. We think that was the case; the direction they thought was north, was in fact, northeast.

On the map in Figure 10–1, there are a series of letters (***a***–***f***) that begin more or less north of the camp area and end northeast of the camp area. They follow a ranch road that extends along a region called the *One O One Flat*. The lights that the two hikers saw during the night of their campout came from vehicles traveling along this road.

In Figure 10–1, the ranch road enters the map from the upper left, proceeds diagonally across the map, and down along the east side of Goat Mountain. The terrain contours show that the road comes through a pass in the hills at location ***a***, and continues down-hill passing the One O One Ranch.

The site of the One O One Ranch is historic. Both a ranch, and the road to it, appeared on the Alpine Texas Sheet of a United States Geological Survey (USGS) map published in 1895.[3] At that time, the ranch was labeled the *McIntire Ranch.*[5] Today, the road going back north from the ranch intersects US90. A century ago, it went straight past the current junction with US90 and continued on to the Paisano Station of the Galveston, Harrisburg and San Antonio Railroad.

Notice that the central region illustrated in Figure 10–1 is labeled *One O One Flat.* Black described this area as the *Marfa*

[5] While the ranch shown on modern USGS maps is now labeled *One O One Ranch*, the McIntire name, spelled *McIntyre* on current maps, is still attached to local geographic features. There is a McIntyre Creek, and McIntyre Peak at 6,378 feet asl is located just off the upper right side Figure 10–1. It is also on Figure 2–1.

flats, and today many people refer to the Marfa plain as the *Marfa flat* or the *Mitchell flat* — using *plain* and *flat* interchangeably. But in the USGS' lexicon, the two are not synonymous. *One O One Flat* was a label assigned by the USGS to this region because of the degree of flatness present in this area of the Marfa plain. (See *flat* in the *Glossary*.) The label *One O One Flat* appeared on USGS maps by at least 1983.

To unravel what Black and his companion might have seen, we need to consider how the terrain between the ranch road and the campsite can impact the visibility of light cast by headlights as a vehicle moves down the road at night.

In Figure 10–1, location ***a*** on the ranch road is at a pass over the northern hills. The elevation there is 5,149 feet. That height is considerably above the region of the One O One Flat where the campers were, which has elevations in the 5,000–5,080 range. A vehicle coming through the pass would have its headlights aimed more or less straight out across the One O One Flat. With the height advantage provided by the pass, headlights could be seen throughout most of the One O One Flat. In particular, the lights could be seen from anywhere inside the camp area at the base of Goat Mountain — provided there was no local vegetation to obstruct the view.

To illustrate that the campers could see auto headlights at ***a***, requires some detailed calculations. Figure 10–2 shows the terrain along a line from ***a*** to the position marked ***X*** (elevation 5,043 feet) on the map in Figure 10–1. Note the difference in the scales of the vertical and horizontal axes. Vertical distances are measured in feet above 5,000,[6] while horizontal distances are measured in miles — a factor of more than 5,000 to 1. It is clear that vehicle lights could shine (essentially horizontally) all the way to the base of Goat Mountain.

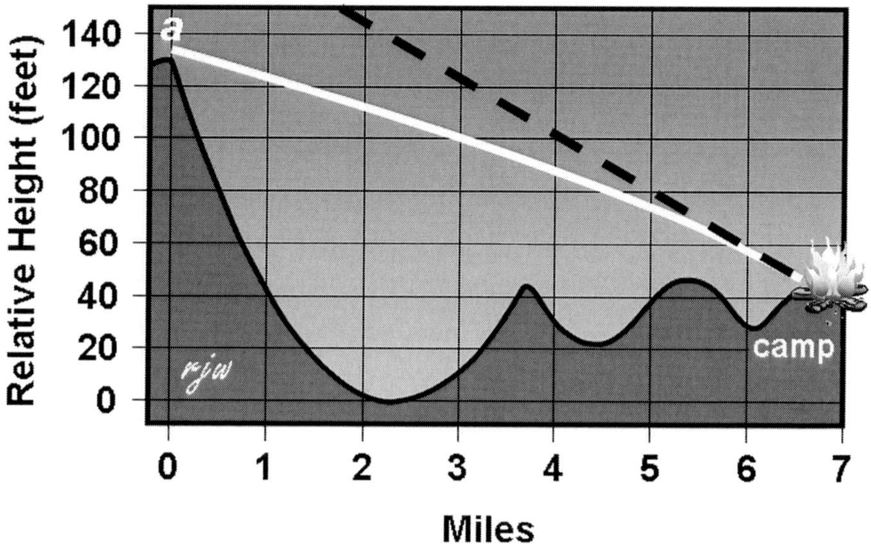

Figure 10–2. The bottom line indicates the terrain from the pass (***a***) north of the One O One Ranch to the possible campsite at the base of Goat Mountain as shown in Figure 10–1. The white line is the path of light cast by headlights at location ***a*** shining across this area to the campsite. The dashed black line is tangent to the white light beam at the campsite. The campers perceive the light as originating in the direction of the dashed black line.

Two little hills show up in Figure 10–2 at miles 3.7 and 5.4. They are parts of two sequences of small hills and undulations in the middle and bottom of the One O One Flat. These hills are also shown on Figure 10–1 inside the upper and lower dashed

---

[6] All terrain heights are shown relative to the 5,000 foot elevation at the latitude and longitude of the campsite. The curvature of the earth has been removed from all heights.

ovals labeled *small hills*.

Light from a truck follows the path indicated by the white line in Figure 10–2 which arcs out over the One O One Flat to the campsite. A small amount of curvature, due to refraction, is clearly evident. From Figure 10–2, it is apparent that the hills do not present any impediment to the light projected by the headlights. In fact, the fan of light rays coming from the headlights can illuminate the entire distance from ***a*** to the campsite.

The dashed black line is the direction the campers would perceive the light to have come from. It is tangent to the white ray of light that arrives at the campsite. This is another illustration of looming as we discussed in *Chapter 10 – Larger Than Life*. Though there was a strong inversion on the One O One Flat in the early morning hours of December 1, 1985,[4] and every light path we analyzed showed some looming, we'll find that refraction played a minor role in the explanation of what the campers saw that night.

The little hills in the center of Figure 10–1 eventually do become a barrier to headlights as the vehicle descends from the northern pass. By the time a truck reaches ***b*** — one-half of the way from the pass down to the One O One Ranch — it has descended to an elevation matching that of the small hills. At that point, its headlights would just graze the tops of the hills. The interplay between road elevations, hills on the One O One Flat, vegetation along the ranch road, magnitude of light refraction and reflection, and exact position of the campsite is what determines whether lights can be seen *from* the campsite.

Figure 10–3 illustrates elevations along the ranch road. The path is labeled from ***a–f*** just as in Figure 10–1. In both Figures 10–1 and 10–3, the segments where vehicle headlights are visible from the campsite are colored white. The first segment extends 1.3 miles from ***a*** to ***b***.

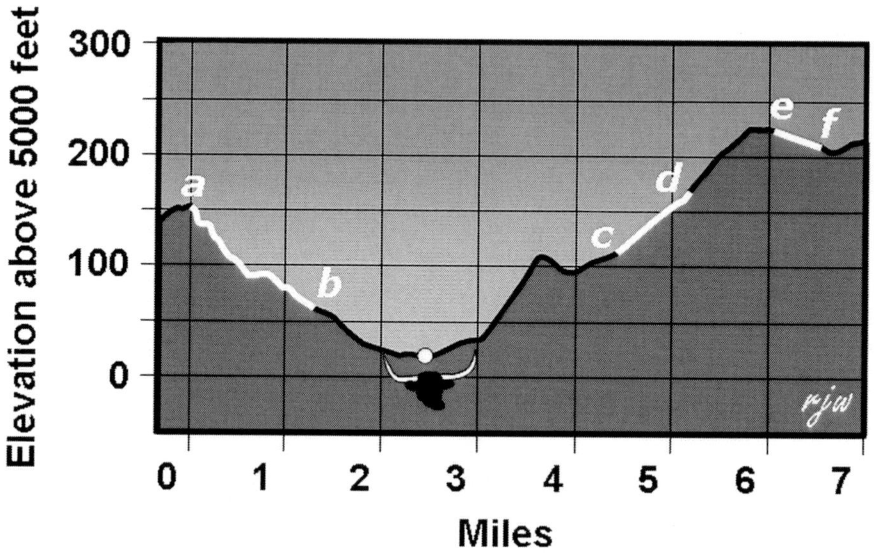

Figure 10–3. Elevations along the ranch road from US90 south to Goat Mountain.

Proceeding south on the ranch road from ***b*** to the One O One Ranch, a vehicle would remain lower than the upper range of small hills and be unable to project its headlights out onto the One O One Flat. After passing the ranch, the road heads to higher elevations. By the time location ***c*** is reached, the elevation is once again high enough to allow headlights to shine out with grazing incidence onto the One O One Flat and onto some positions within the camp area shown on Figure 10–1. The road then continues to ***d*** where the lower group of small hills in Figure 10–1, begins to block headlights from reaching the campsite.

Throughout the 0.9 miles from ***c*** to ***d***, visibility of headlights is strongly dependent on exactly where the campsite is located. The location determines where a ray of light would pass through the lower range of hills/undulations shown on Figure 10–1. The low spots in the undulations are not so deep that they could be called *passes*, but the high spots can definitely block headlights.

The last segment along the ranch road that is indicated as having visible lights, ***e–f***, is about a half mile long.

This lengthy introduction to the One O One Flat and the roads, hills, and camp area now enables us to examine the campers' description in more detail to see how the unique conditions of the One O One Flat produced what Black and his friend saw.

Black's statement that the lights always reappeared in the same location indicates that the men were not seeing some atmospheric apparition that is blown about by the winds. Nor was it like the blooming of a flower at one spot, and then at another. The appearance of the lights was correlated to a specific location on the earth — such as a road. Although there are several places along the ranch road where lights could reappear or extinguish, we believe that locations ***a*** and ***c*** are the most probable candidates.

If the initial sighting occurred at ***a***, the lights would have been visible throughout the entirety of the 1.3-mile long segment ***a–b***. At 45 mph, a vehicle would cover that distance in 1.75 minutes, which is a lot longer than Black's estimate of a few seconds. Either his time estimate was faulty, or Black and his friend didn't see lights from this segment — perhaps because vegetation blocked the view from their campsite.

The second segment from ***c–d*** does not provide continuous visibility of light coming ***directly*** from a vehicle's headlights for two reasons: (1) Small hills near the campsite obstruct the view of light coming from the northern end of this segment. (2) Near the middle of the segment, the angle of the road is such that the headlights just don't point toward the campers.

But what if the light that Black and his friend saw was being reflected somehow?

We first examined the possibility that the headlights were illuminating fog. However, the weather conditions were not conducive to the formation of fog that night. Even though it didn't seem likely, we considered whether the road surface itself could provide enough reflected light. Bingo! In-depth mathematical analysis indicated that reflected light could be seen by the campers. But reflected from exactly what was the question?

Light from the middle of segment ***c–d*** could be visible for a few seconds at a time as headlights illuminated the roadbed, the terrain on the sides of the road, any plowed edges beside the road, vegetation along the side of the road, and **especially any lack of flatness on the road itself**. The campers would not be seeing the lights directly, but instead, would see illuminated surfaces in front of the vehicle.

Previously no one has described Marfa lights as being reflections of lights shining on the ground. It's likely that nearly all Marfa lights shine directly from the source to the viewer without intermediate reflection. What makes the current situation possible is that the campers were less than three miles away from segment ***c–d.*** If you turn to Table 5–2 in *Chapter 5 – Illuminating the Mystery*, you'll see that light from a single candle can be seen from three miles away in full moon light. So it wouldn't take much reflection to be strong enough to be seen.

Now, consider that Black said the lights looked like "a horizontal length of light" and "little beams of lights". Well, the headlights on the front of a truck traveling on an undulating road would project out, illuminating a beam-shaped swath on the earth. The reflection from that plume of light would appear as a horizontal length of light when seen from the campsite. This arrangement of road, headlight, and reflection is shown in Figure 10–4.

Figure 10–4. Idealization of the road south over segment ***c–d***. Cathedral Mountain appears on the right-hand side of the figure, and the ranch road curves off to the east toward the mountain. The road generally slopes uphill. But the slope at any specific position can be more or less than the average slope. As the truck follows the undulations of slope, the beam from its headlights rocks up and down.

The headlights of the truck on the left-hand side of the figure are shown projecting a light beam that illuminates the road in front of the vehicle. As the truck encounters the undulating features on the road, the beam goes up and down. The headlights "catch" the up slopes and brightly illuminate them but "miss" the down slopes, leaving them in the dark. Similarly, to the sides of the road, contours of the terrain and any adjacent vegetation are illuminated on the side facing the light.

Black stated that the lights danced up and down and vibrated. This is the light behavior you'd get from a truck driving on a road with varying slopes like the uphill segment from ***c*** to ***d***. As the truck heads toward ***d***, the swath of earth illuminated by its lights is either higher or lower depending on the slope of the road at the truck. Witnessing this from three miles away, our campers would see a beam of light that seemed to "dance" up and down. What Black and his friend were witnessing was the reflection of headlights on the road and surrounding terrain. That is why the light appeared as a beam and not a spot of light.

Black also made the general observation that the lights moved across the distance in front of them. Regrettably, Brueske's account of their story doesn't say whether the lights moved to the right or to the left, though we believe they moved to the right.

All three light segments in Figure 10–1 would produce light that moved across the segment. The angular sweep across segment ***a–b*** is the smallest of the three. It is about 2.5° from the pass at ***a*** until a vehicle's headlights slip behind the small hills at ***b***. This angular change may seem small but it is well within a person's perception. To get a feeling for what the campers would have seen, have someone stand 20 feet away from you and move a flashlight one foot from left to right. We think that you would describe the light beam as "moving across".

Seen from the vantage point of the campsite, the total angular change of a vehicle moving from ***c*** to ***d*** is really big — greater than 30°. If lights were visible throughout this entire range someone would certainly describe them as "moving across".

From location ***d*** on over to ***e***, the lower range of small hills again blocks visibility of vehicle lights toward the campsite. Within that span of the ranch road, there is a junction with one branch leading to Cathedral Mountain and the other branch continuing down east of Goat Mountain. At the junction, the branch leading behind Goat Mountain turns sharply toward the campsite. By the time the road gets to ***e***, and is above the small hills that had been blocking light to the campers, the vehicle is pointed to within 64° of the campers. The **direct** light of the truck's headlights would be very bright at such a close range (2.6 miles).

The campers would notice the contrast between the last weak light they saw at ***d*** (which came from light **reflected** off the road, adjacent terrain, and vegetation) and the bright light now coming **directly** from the headlights at ***e***. They would also react to the new apparent height of the light above them. The increased elevation of the road plus refraction produce a perceived light at the campsite that is a full one-quarter degree higher in the sky than when it was last seen at location ***d***. That's an angular change equal to half the width of a full moon. **And, someone would rightly refer to that as seeming "to shoot straight up vertically"!**

Black indicated that the lights came back down, danced horizontally, and then disappeared. As the lights moved from location ***e*** to ***f***, they did move horizontally with an angular change from left to right of ~10°, but they didn't go back down.

In that segment, the lights stayed at about the same apparent angle up in the sky. But the horizon behind the lights, which is defined by Cathedral Mountain, moved upward. With the increasing elevation of the horizon behind the lights, Black and his friend were tricked into thinking the lights went down. The lights ultimately disappeared when, once again, the small hills in the lower range of hills got in the way.

Reports of lights on the Marfa plain that shoot straight up are not common. In fact, they are rare. This rarity was the primary reason we chose to investigate the sightings on the One O One Flat. In this case, there is a very plausible explanation for the camper's perception. We'll examine another report of lights that "go up" in *Chapter 11 – Aragon ~ Lights Below the Chinatis*, and find entirely different conditions, facts, and conclusions.

But what about Black's description that the lights ranged from yellows and oranges through reds? This leads us to quite a different line of inquiry from any up to now. We need to know what influences the color we see in a light display.

Above we indicated that the campers saw reflected light coming from the terrain and road that are shown in the map in Figure 10–1. What color should the reflected light be? Well, that's a bit of a trick question, because you need more information in order to answer it.

Instead, let's ask what color is the terrain in the daylight, say early morning or late afternoon? You can answer that with confidence because you can just go look at it. We'd say the soil is light brown or tan. But that's not what color the soil is at night. Go look at the soil some night when the moon is not showing. The soil is very dark, tending toward black.

The answer to the above rhetorical questions, and the fact that the "color" of the soil is different from day to night, is that soil doesn't have a "color" of its own. It gets much of the color we perceive from the light that shines on it.

During the day, the light that shines on the earth comes from the sun and has a "color" that we casually call *white*. The light we receive is well described as coming from an object with a temperature of ~5,500°C. Such an object emits light that has a broad peak of energy right in the range of our vision. This broad range of "colors" falls on objects on earth — such as the soil of the Marfa plain — and is reflected. But the reflection is not uniform. The soil reflects some colors more than others. The collection of reflected wavelengths of light produces the composite perception that we call the color of the soil.

On the night of the campers' observations, a very different "coloring" of the soil occurred. In the first place, the soil was illuminated with headlights from a vehicle. Headlights do not produce a "white" light like the sun does. Instead, the spectrum is very much more yellow-red. We can quantify what it looks like.

In the 1980s, all three of the major automobile manufacturers used the same headlamp in their pickup trucks[7] — the H6054 (Sylvania version).[5] Their passenger cars used this headlamp also. But let's focus on the pickup trucks because they are more likely to have been traveling along the road that night.

Sylvania specifies that the lamp had a color temperature of

7 The Sylvania H6054 headlamp was used during this period on the following full-sized pickups: Chevrolet from 1979–1987, Dodge from 1983–1987, and Ford from 1979–1986.

~2,900°C. That means that the headlamp emits light as though it came from a filament with a temperature of ~2,900°C. Throughout the visibility range of humans, this headlamp produces a spectrum starting with a quite weak blue light that steadily intensifies as the wavelength increases to green, yellow, orange, and red. The peak of radiation produced by this headlamp is well out in the infrared region, invisible to humans. These facts show that just accounting for the headlamp, leads to a light that is more yellow-red in appearance than sunlight.

But there is more to the story. When the light strikes the soil, it reflects. And it does so in a way that biases the light to an even more yellow-red hue. The soil throughout the One O One Flat (and the Marfa plain generally) is termed *Reagan silty clay loam.*[6] The Reagan series of loam is one of the most common groups of soils found in the entire Trans–Pecos region. They are found in about 13% of the area. The official description of *Reagan silty clay loam* says that the soil is brown or grayish-brown. But it is noted that if you look below the pristine surface, the soil becomes light brown, light yellow, or buff. Of course, trucks driving on every ranch road would disturb the surface and reveal subsurface soil. Running a road grader over the road would cut down even deeper into the soil. The official description of the soil, and its appearance to any viewer today, would certainly make one inclined to think that its reflection would produce yellow-red appearing light.

But we can go still further in determining the appearance of light reflections. It is generally known that dry loam soils have a reflectance that is low for blue light (~10%) and relatively larger for red light (~30%). The variation in reflectance for wavelengths between blue and red is close to linear. We can insert this form of reflection, and the spectrum of a 2,900°C headlamp into an analysis system[8] and calculate the color that results. We found the resultant light, when expressed in "intensities" of the primary colors to be, 47 for red, 41 for green, and 12 for blue. When this is plotted on a standard color map such as the one in Reference [7], it is solidly in the region one would call yellow-red.

What we've described above treats the reflection process as though there is only one composition of soil, and only one pattern of reflection. In reality, the undisturbed soil along the path of the road would vary in appearance. Grading, wear, and erosion would reveal variation in the soil's appearance on the roadbed, also. Together these variations would make the perceived color of the light beam fluctuate as the truck pitched along to its destination.

There is yet one more dimension to the coloring of the beam of light seen by the geologist and his friend. It has to do with the haze and humidity on the Marfa plain on the night of the men's camp out.

Hazy skies are so common — particularly in urban areas — that everyone has seen them. But, people may not associate haze with isolated areas away from metropolitan centers. In fact, haze is prevalent in the desert southwest.

In the daytime, haze is likely to present a color that is a blend of white, light gray, and pale blue. This color is caused

---

8 The *Commission Internationale de l'Eclairage* (CIE) developed a standard for displaying light composed of the three primary colors in 1931. Called the *CIE 1931 XYZ Color Space,*[7] it enables a color to be specified mathematically and illustrated by a specific colored example.

when air molecules and particulates (haze) in the air scatter sunlight back to our eyes from all locations along the direction we're gazing. The scattered light obscures the view of distant mountains.

The effect produced by haze is evident in Figure 10–5.

Figure 10–5. View from the parking lot at the University of Texas McDonald Observatory on Mount Locke at 6,778 feet asl taken on December 9, 2011. In the center, Cathedral Mountain — 6,785 feet asl — is 40 miles away. Its bearing from Mount Locke is 148°. At 6,883 feet asl, Twin Mountains bracket Cathedral Mountain. They are 21 miles from Mount Locke and about 10 miles north of the MLVC. Our intrepid campers spent the night on the One O One Flat just a few miles from the right shoulder of Cathedral Mountain.

This photograph was taken from the parking lot at the University of Texas McDonald Observatory on Mount Locke looking approximately southeast on a day when the winds had stopped. Notice that the tops of the mountains sticking up above the haze are relatively distinct. But at lower elevations, the haze increasingly obscures the features of the landscape as they become more distant from the viewer.

Although the haze has the same density throughout the region, the more distant an object is from the observer, the more blue/white haze the observer looks through, and the less visible the object's details become. This is a general consequence of haze during daytime. As we look at a distant object, light coming from the sun and clouds strikes particles in the haze between us and the object, and the light is scattered in our viewing direction.

In the evening, the effect is just the opposite of that seen in Figure 10–5. Now, as we look in the **direction** of the sunset, not just haze, but the entire atmosphere scatters blue light out of the sunlight before it reaches our eyes. When the white light of sunshine has the blue color removed, all we can see are the remaining yellows, oranges, and reds. This yields the beautiful sunsets common in the desert southwest.

Exactly which colors get scattered out of a light beam and which ones remain behind is determined by the sizes of the particles in the haze. Particles with diameters somewhat larger than the wavelengths from blue to red scatter all colors equally. Particles with small diameters — say, comparable to the wavelength of blue light — scatter blue light more than red. And, particles with very small diameters don't scatter any of the light that we are able to see. Generally, the very small scatterers are the most abundant.

Colorful sunsets are just one example of the effect of haze. Lights from vehicles on the One O One Flat can also be scattered as they shine through haze. So we would expect some degree of reduction in the blue component of light from headlamps. Like the sunset, the colors that we perceive from headlights are dependent on the size of the haze particles in the local air. And, particulate size is enhanced by yet another factor — humidity.

It is very well know that high humidity can lead to the accumulation of water molecules on the surface of haze particles. At a humidity of 80% or higher, the effective particle size can more than double. This means that a dry particle of haze that was too small to scatter blue light effectively can grow under high humidity conditions to a diameter that makes it thoroughly effective at removing the blue light.[9] Indeed, that is likely to have been the case starting in the early morning hours of December 1, 1985, because the Marfa Airport reported humidity values at 80% or more from about 1 a.m. until sunrise.

Black reported seeing lights "a little after midnight". But without a watch, we suspect that he was off in his estimate of the time. It is likely that his observations didn't occur until even later than that — after the humidity on the One O One Flat reached higher levels.

We calculated above that the light reflected from the roadbed and adjacent terrain only had an intensity of 12 for blue hues. Now, if the humid haze scatters enough blue light out of the beam to bring the blue content down to a reading of 7, it would make the remaining light beam distinctly orange in appearance.

---

[9] In Reference [8], the researchers found that humidity in the Big Bend was a significant factor in reducing visibility. Fine sulfate particles that absorbed water molecules were the primary cause of visibility reduction.

We will never know quantitatively what role haze and humidity played in the appearance of the light beams the campers saw. But it's safe to assume that the color of the beam was produced by a combination of three factors — color content projected by the headlamps, wavelength-dependent scattering of light by the soil, and wavelength-selective scattering of light out of the beam by particles of haze in the air.

This still leaves us with one more question. Who would be driving along a ranch road during the early morning hours of December 1st? The obvious answer is ranch personnel. The initial traffic after midnight could have been ranch hands returning from a Saturday night out. And, later Sunday morning, it could also have been ranch hands who wanted to get onto the range early. But a collection of four to five lights at approximately one-hour intervals still seems like a lot of nighttime ranch traffic.

Who else might it have been? It turns out there ***was*** another group of people that might have been present — hunters.

December 1, 1985 was the second day of mule-deer hunting season in Brewster County. Deer hunting in Brewster County is big business. The Texas Parks and Wildlife Department estimates that during the 1985 season, in the Trans–Pecos Wildlife District (which includes Brewster County), about 8,300 hunters took to the field. They harvested over 8,000 mule deer. Though some hunters are likely to stay at the ranch that hosts their hunt, others stay at local motels and drive to the hunting site each morning. In order to be in position by sunrise, they get under way early. Little did they know that their pursuit would also make them the architects of this mystery.

It's quite amazing how a collection of humdrum facts can fit together to make a coherent explanation of a seemingly inexplicable experience. In this chapter, the explanation required the campers' detailed story, micro-detailed knowledge of the local terrain and roads, the physics of light's spectral content, the physics of soil reflection, the physics of scattering by hydrated particles of haze, and the hunting season in Brewster County, Texas. All of that distills down to —

> Some hunters were driving along an undulating road south of the One O One Ranch in the early morning hours of December 1, 1985. Their headlights, reflected off the roadway in front of them and encountered haze that further reddened the light beams that were then observed by two campers at the base of Goat Mountain.

It takes just two sentences to sum up the mystery now. But it took years to put all the pieces together!

The next two chapters present more observations that have equally obscure explanations — lights that "go up" and multicolored lights that chase cars. Can you imagine how that's possible?

— § —

10 — Sources

[1] Judith M. Brueske, *The Marfa Lights — Being a collection of first-hand accounts by people who have seen the lights close-up or in unusual circumstances, and related material,* revised ed. (Alpine, TX: Ocotillo Enterprises, 1989), 16-18.

[2] District Traffics Maps 1936-1990 [CD], (Austin, TX: Texas Department of Transportation, 2004).

[3] USGS Map Product 301227, "Alpine, TX Historical Map Geopdf 30x60 Grid 125000-scale 1895." Available from: http://store.usgs.gov/b2c_usgs/catalog/query/(ctype=areaDetails&xcm=r3standardpitrex_prd&carea=%24ROOT&layout=6_1_61_48&uiarea=2)/.do?query=301227&lastVisited=areaDetails

[4] NOAA Earth System Research Laboratory, Physical Sciences Division, "3-Hourly NCEP North American Regional Reanalysis (NARR) Composites," NOAA. Available from: http://www.esrl.noaa.gov/psd//cgi-bin/data/narr/plothour.pl

[5] Sylvania "OSRAM Sylvania Products, Part Number H6054." Available from: http://www.sylvania.com/en-us/products/Pages/default.aspx

[6] W.T. Carter, M.W. Beck, and H.M. Smith (United States Department of Agriculture), and H.W. Hawker, E.H. Templin, and T.C. Reitch (Texas Agricultural Experiment Station), "Soil Survey (Reconnaissance) of the Trans–Pecos Area, Texas." *Bureau of Chemistry and Soils, United States Department of Agriculture, Series No. 1928, No. 35.* (Washington, D.C.: U.S. Government Printing Office, 1934).

[7] Thomas Smith and John Guild, "The C.I.E. colorimetric standards and their use," *Transactions of the Optical Society* 33, no. 3 (1931): 73-134.

[8] W.C. Malm, D.E. Day, S.M. Kreidenweis, J.L. Collett, and T. Lee, "Humidity-dependent Optical Properties of Fine Particles During the Big Bend Regional Aerosol and Visibility Observational Study," *Journal of Geophysical Research* 108, D9 (2003): 4,279.

11 — Aragon ~ Lights Below the Chinatis

TRAVELERS ON US90 can easily miss Aragon as they speed by at 75 mph. There are none of the ghostly remnants of an old town; no relics of buildings or old roads are visible from the highway.

Two shiny metal sheds bearing the word *Aragon* stand as mute sentinels alongside the Southern Pacific Railroad tracks, reminding the traveler that the community of Aragon once stood there. The shed nearer to Marfa — 9.1 miles northwest of the intersection of US90 and TX17 in Marfa — marks the approximate location of this early 1900s community.

To see definitive markers of Aragon's former location, you

must take to the air. From aloft, scars left by the runways of a U.S. Army World War II auxiliary airfield, are readily visible. The Aragon airfield, which was deactivated in 1945, is just south of US90 across the road from the Aragon shed labeled, *CPSA 643*.

Today, where Aragon once stood, there is grass extending along both sides of US90 for as far as the eye can see. In the spring, the fields are green; and by summer, they turn golden brown like wheat. There are places near the horizon where mirages shimmer above the grass blurring the bottom of the distant mountains. The grass can easily be knee deep.

But these depths are modest compared to what existed when the cattlemen settled here in the late 1880s. For example, in his unpublished memoirs, Robert Reed Ellison wrote about his trip across the Marfa plain —

> *When we bedded our cattle down that night we saw no more of them until the next morning as the grass was so high and soap weed*[1] [sic] *so thick they were hidden from view.*[2]

Now that's a lot of grass![3] Could grass have played a role in

---

1 *Soapweed* is another name for the *Yucca elata* plant which grows throughout the Trans–Pecos, mostly at elevations above 4,000 feet. The crushed roots of the soapweed were used to make soap. The stems of some plants can reach nearly 30 feet high.[1]

2 Cecilia Thompson's Reference [2] contains the quote from Robert Ellison.

3 There are three major grasses on the highland plains connecting Valentine, Marfa, Alpine, and Fort Davis. They are the Blue Grama, Black Grama, and Sideoats Grama. Blue Grama is the dominant grass. Both Blue and Black Grama grow up to two feet in height. In contrast,

the naming of Aragon?

The history of the Southern Pacific's El Paso Division[3] doesn't mention how Aragon got its name. From that we can assume, Aragon was not named after a Southern Pacific employee. Nor was the name derived from the Mexican and Indian cultures. So, where did the name *Aragon* come from?

We know that in early 1883 Chief Engineer James Strobridge inspected construction of the railroad east of El Paso, and that his wife, Hannah Marie, suggested the name *Marfa* for the watering stop on the banks of Alamito Creek. (See *Chapter 2 – The Marfa Plain.*) *The Galveston Daily News*[4] indicated that other sites besides Marfa were named after characters in Jules Verne's *Michael Strogoff.* Yet, Marfa was the only name the authors could find among the Jules Verne characters.

Is it possible that Hannah Strobridge also suggested the names for other railroad sites — like Aragon? If so, what might have inspired Mrs. Strobridge to select the name *Aragon* for this particular stop?

The answer may lie in the grass. The abundant, tall grass of this area was very reminiscent of the wheat fields of another Aragon — a province in Spain. Situated on the French border with the northern edge of the province in the high peaks of the Pyrenees Mountains, Aragon falls to lower elevations as the terrain gets further from the French–Spanish border south of the Pyrenees. There, the relatively flat lands extend for miles and are covered by golden wheat fields.

The view across the wheat fields toward peaks and lower

Sideoats Grama grass can reach four feet in height. It is the state grass of Texas.

mountains of the Pyrenees is similar to the view from the old Southern Pacific Railroad tracks toward the Davis Mountains. In addition to the visual similarity, a viewer familiar with the abundant wheat production of Spain's Aragon region would have been struck by the magnitude of the grass in the Marfa–Valentine basin. By all appearances, Aragon, Texas was an image of Aragon, Spain.

It may seem unlikely that anyone in Texas knew about the wheat fields of Aragon. But Aragon, Spain was front-page news in Galveston in 1879 when floods produced widespread damage and washed out major railroad infrastructure.[5] Even before that, the significance of the Aragon, Spain wheat fields was well documented.

Writing in his five volume treatise in 1809, Frenchman Alexander De Laborde states —

> *Spain produces a considerable quantity of wheat. It is grown in all the provinces of this monarchy in a greater or less proportion. ... **Aragon grows also a larger quantity than it consumes**, and with the surplus satisfies the demands of the neighbouring provinces.*[6]

Aragon's wheat production was also featured in European travel books during the late 19th century. In 1865, Bradshaw published a travel guide to Spain and Portugal. New editions were published *in English* each year until 1895.[7] So, Hannah Marie Strobridge could easily have obtained a current copy of the guide.

In the 1867 travel guide, Bradshaw when writing about the town of Tortosa in the province between Aragon and the Mediterranean Sea said —

> *It has important trade through its two ports . . . in the* ***wheat of Aragon****, timber from the Pyrenees, Aragon, and Catalonia; . . . .*[7]

Although there is no way to substantiate how and why the railroad stops were named without finding a document such as the diary of Hannah Marie Strobridge, the circumstantial evidence that Aragon, Texas was named for its similarity to Aragon, Spain is compelling.

While the Aragon in Spain has the Pyrenees Mountains, the Aragon in Texas has the Davis Mountains ***and mysterious lights***. But these are not the usual mysterious lights that move sideways. Not at all. Aragon's mysterious lights have been seen to move upward! A boy, named Jack Reed, provides our story.

Jack Reed was born in Marfa in April of 1927. He lived in the Big Bend area most of his life. More importantly, sometime around 1939, he visited his cousin in Aragon. At night, their discussions turned to mysterious lights, and Jack's cousin pointed out the lights on the Chinatis that were visible from their house in Aragon. Many years later, Jack recounted his experience to Judith Brueske —

> *"This was at age about eleven or twelve," says Reed. "We were sitting on the porch.. and you could see the lights...And we discussed that maybe the lights were where a road goes down there." But his cousin said that they had already asked about that possibility and concluded that "It's not lights on the road because, more or less, they go up rather than sideways."*[8]

While we have read various accounts of lights going up, this is the only eye witness account suggesting that the phenomenon was repeatable. We had to answer the question —

Why do the mysterious lights of Aragon move upward? But before we can do that, we'll need to understand more about the environment around Aragon. Let's begin with the topography.

Even with tall mountains visible in the distance, travelers passing by the site of Aragon on US90 today might not appreciate that they — and Aragon, Texas — are in a bit of a valley. Looking toward the northeast and southwest of US90, you can see that the terrain rises toward the horizons. But the distances are so large that the heights seem deceptively small, until mountains shoot up abruptly.

The general horizon to the southwest has an elevation over 900 feet higher than that of Aragon, which is 4,900 feet above sea level (asl). Rising 7,730 feet asl, the highest prominence in the southern view is Chinati Peak — 33 miles southwest of Aragon. North of Aragon, many of the peaks of the Davis Mountains surpass Aragon's elevation by 3,000 feet or more. Twenty miles due north of Aragon, the highest of these — Baldy Peak — is 8,378 feet asl.

The relation of all these features to one another is illustrated in Figure 11–1. Baldy Peak, Chinati Peak, and all the many black dots on the map represent named mountain peaks with elevations above 6,000 feet.

US90 runs from the northwest through the valley between the Davis Mountains and the mountains bordering the Rio Grande to the southwest. Along the course of the highway, five locations came into existence due to the creation of the Southern Pacific Railroad in the 1880s. They are Valentine, Quebec, Aragon, Marfa, and Alpine. Three of these became small population centers. Today, Quebec and Aragon are just railroad sidings.

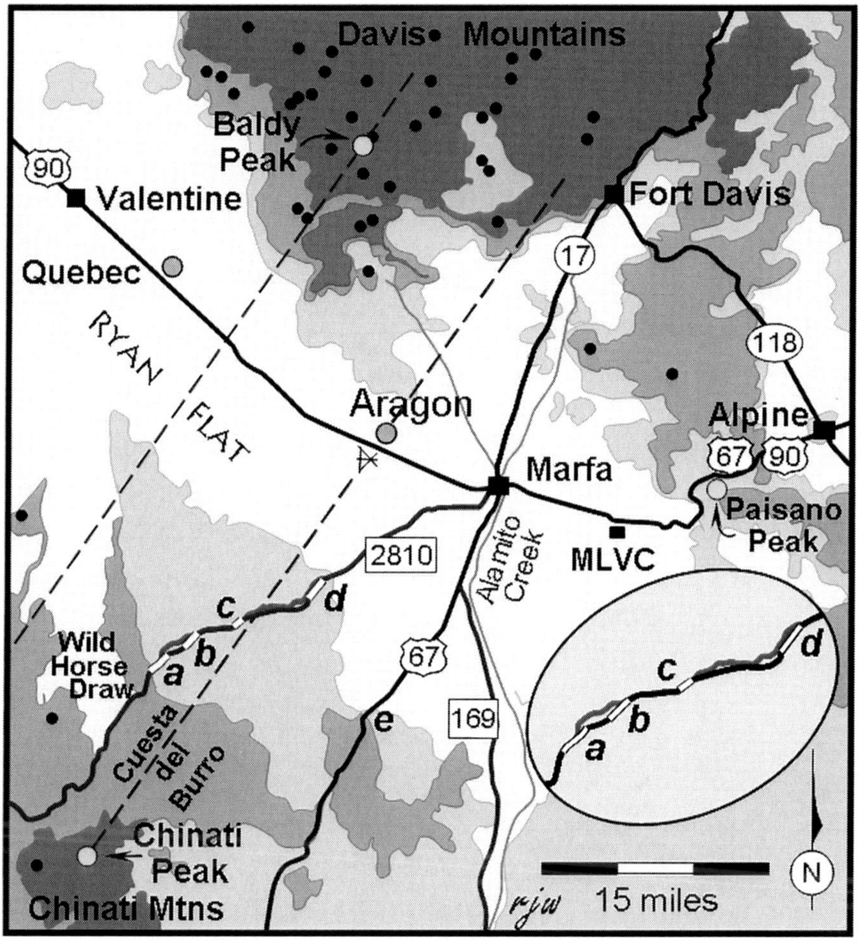

Figure 11–1. Map of the region surrounding Aragon, Texas. Black dots indicate mountain peaks with elevations above 6,000 feet. Road segments ***a***, ***b***, ***c***, and ***d*** (magnified in the inset) have true bearings from Aragon of 224°, 223°, 217°, and 202°. (See *bearing* and *magnetic declination* in the *Glossary* for a discussion of *true bearing* and *magnetic bearing*.)

Driving toward Marfa along US90, if you look to the left and right of the road ahead of you, toward the distant horizon, you can regularly see mirages above the dry grass in the vicinity of the Quebec siding.[4] The mirages can be recognized by the fact that the end of the foothills seem to float in the air with a strip of sky beneath them. Mirages have been seen in this region for nearly a century. The 1925 history published by the El Paso Division of the Southern Pacific Railroad mentions mirages south of the railroad and east of Valentine (in the vicinity of Quebec). It included the following description — "Frequently the ranch buildings and trees at some distance from the track appear to be floating in water."[9]

Heading south-southwest out of Marfa, US67 ascends the foothills adjacent to the Marfa plain on its path toward Presidio. Locations along US67, in the foothills south of Marfa, are where the most commonly reported Marfa lights originate. These lights, which are only about 20 miles from Aragon, were described in detail in the *Chapter 6 – "Classic" Marfa Lights.*

With few exceptions, US67 follows the same path today as it did in 1939. As the authors can attest, some of the lights on US67 that are visible from the Marfa Lights Viewing Center (MLVC) are also visible — without binoculars — from the Aragon siding. Only at Aragon, they are dimmer because the road points more toward the MLVC than toward Aragon. With binoculars, car lights on US67 are very evident from Aragon.[5]

Jack Reed should have been able to see the lights on US67 even more easily than we can today because in 1939 the path of the highway at location ***e*** in Figure 11–1 was aimed more

---

4 Metal sheds beside the Southern Pacific Railroad indicate the location of the Quebec siding which is 28 miles from the intersection of US90 and TX17 in Marfa.

5 The bearing of the classic Marfa lights as seen from Aragon is 182° from true north. See *magnetic declination* in the *Glossary* for a discussion of *true bearing* and *magnetic bearing*.

directly toward Aragon.[6] While it's possible that Jack saw lights from US67, these lights would have been weak and only been visible briefly from his vantage point — making it less likely they were the source of his mysterious lights. We needed to look for a source other than the classic Marfa lights on US67 as the explanation for what the boys saw.

Also leaving Marfa is Texas RM2810. Bearing southwest from Marfa, it passes through Pinto Canyon, and eventually reaches Ruidosa on the banks of the Rio Grande. As the road climbs steadily out of the Marfa–Aragon valley and up the slopes of Cuesta[7] del Burro, it reaches an elevation of 5,893 feet asl before descending precipitously into Wild Horse Draw west of the highlands of Cuesta del Burro. At its peak elevation before the decent into Wild Horse Draw, RM2810 is nearly 1,000 feet higher than Aragon.

Drive out RM2810 to section ***d*** in Figure 11–1 (15 miles from the intersection of Highland Street and US90). Here you can see a panorama of this entire region. Along the way, you'll notice that at about 13.5 miles, RM2810 climbs a steep hill of 150 feet or so, moving you distinctly up from the valley onto the lower slopes of Cuesta del Burro. Looking north from the center of section ***d***, the Aragon valley, US90, and the Davis Mountains are laid out in awesome display. And to the east, the vista is of the Marfa plain and familiar mountains like Cathedral and Paisano Peak.

The valley and the slopes encountered on RM2810 are best

---

[6] Location ***e*** on Figure 11–1 is the same as the northern end of section ***b*** in Figure 6–1.

[7] *Cuesta* is a ridge with a gentle slope on one side and a steep face on the other.

illustrated by a cross-section drawing of the terrain. On the map in Figure 11–1, the dashed lines indicate two paths for cross sections. One line passes through Chinati Peak and Aragon. The second line — about 12 miles distant — runs parallel to the first one and passes through Baldy Peak. Both paths are illustrated in Figure 11–2.

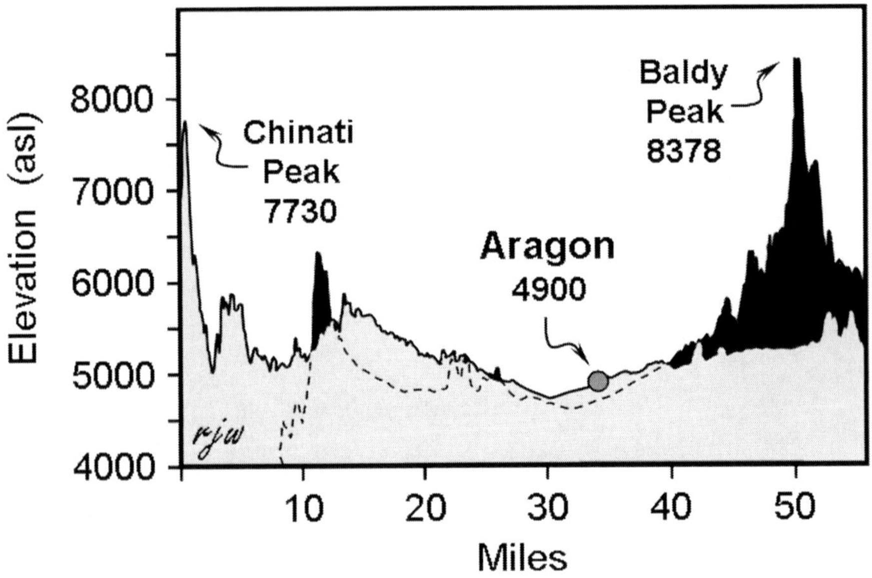

Figure 11–2. Cross sections of the terrain around Aragon showing the shape of the basin. The black cross section shows the terrain profile indicated by the upper left dashed line in Figure 11–1, while the gray terrain follows the lower right dashed line. Notice that the valley floor of the black cross section northwest of Aragon is lower than that of the gray cross section.

You can see from the cross sections that, as US90 and the valley proceed northwest toward Valentine, the floor of the valley decreases in elevation, while the mountainous sides extend upward. This is the setting for Jack Reed's observation of mysterious lights at Aragon.

In his story, Jack compares the relative movements of the lights, saying that mostly they moved up instead of sideways. That's unique. Having a witness make assertions about the comparative movements of the lights, and imply that the display is repeatable makes this a more credible account than the mythological reports frequently encountered.

Further, Jack acknowledges that he and his cousin knew of a road in the direction of the lights, but dismissed it as the possible source since the behavior of the lights did not meet their expectations of how car lights should appear.

Explaining how lights could originate from the direction of the Chinati Mountains, and why they would not appear to move left or right is relatively easy. But explaining how they would commonly seem to move upward is difficult indeed. So let's get started.

Let's return to RM2810 and the map in Figure 11–1. Notice that from the perspective of a person at Aragon, a car heading along RM2810 toward Marfa generally moves from right to left. Also notice that for much of the distance through the foothills below Cuesta del Burro, RM2810 angles off toward the east and does not point in the direction of Aragon. If you were at Aragon today, you could not see lights from vehicles on these easterly-pointing sections of road because the headlights project light only weakly toward the side.

However, at four locations, the road bends to point almost directly toward Aragon. At the positions marked ***a***, ***b***, ***c***, and ***d***, the road is 19.9, 17.8, 14.6, and 11.2 miles respectively from Aragon. All of these locations are close enough that vehicle lights could be seen in Aragon. (See *Chapter 5 – Illuminating the Mystery*). The relatively straight sections of RM2810 at ***a***, ***b***,

c, and ***d*** are 1.4, 1.2, 1.2, and 1.75 miles long.

Because high foothills between ***c*** and Aragon block the light, ***c*** is not a possible location for projecting light to Aragon. The other three locations, however, all have a direct line-of-sight with Aragon. They are the source for the mysterious lights that can be seen from Aragon today. Having witnessed these lights from the Aragon siding, the authors can confirm that they are very bright. (See the caption for Figure 11–1 for the bearings to those road segments.)

But vehicle lights on the current RM2810 are definitely not the source of the lights that Jack Reed saw. His lights came from the original Pinto Canyon Road that ran between Marfa and Ruidosa.[8] That road followed a creek bed in a valley to the west of the current RM2810. Old maps indicate the general path of the old road, but not with the detail that RM2810 is shown on modern U.S. Geological Survey maps. The Old Pinto Canyon Road differed in its path from that of the current RM2810 by distances ranging up to half a mile. In the map of Figure 11–1, the old road's approximate location is shown by the gray line just west of RM2810. This area is magnified in the inset.

We'll begin by describing in detail the lights from the current RM2810 as if Jack Reed were seeing them from Aragon

---

[8] A person might think that the Old Pinto Canyon road was not traveled enough to be the source of the lights Jack saw. But Texas Highway Department traffic survey data of 1936–1938 show that it carried more than a third of the number of autos traveling the portion of US67 that produced the classic Marfa lights.[10] We know from *Chapter 6 – "Classic" Marfa Lights* that even in 1928 the cars producing the lights on US67 were sufficiently numerous to create a popular viewing location which later became the MLVC.

today. Then, we'll use those findings to examine the source of light from the Old Pinto Canyon Road in the shallow valley to the west.

The fact that the three stretches of road, ***a***, ***b***, and ***d***, point toward Aragon explains why the lights are not perceived as moving right-to-left. When the lights are visible, they are moving straight toward a viewer in Aragon. When the road turns and moves toward the left of a viewer in Aragon, the lights seem to "go out" because the headlights project substantially less light toward the side. Headlights project more light toward the passenger side of a vehicle than they do toward the driver's side (the side toward Aragon).

When the vehicle lights on RM2810 point toward Aragon, they are only visible for the time that it takes the vehicle to travel over the segments ***a***, ***b***, and ***d***. Assuming that the vehicle is traveling at today's legal speed limit of 55 mph, that would lead to "lights-on" times of 1.5, 1.3, and 1.9 minutes at ***a***, ***b***, and ***d***, respectively.

Did Jack Reed mean that during a single "lights-on" time, the lights appeared to move upward? If he did, it would mean that in an observation time of about one to two minutes at today's speeds,[9] the lights would appear to move upward a discernable distance. That would be phenomenal indeed.

If Jack was describing a single "lights-on" episode, then the first thing we need to examine is what the road does over those stretches that point toward Aragon. In fact, in each case, the road moves steadily downward. For example, if the light were coming from position ***d***, it would appear to move downward 25

[9] In 1939, the creek bed would have been substantially rougher and the speed of travel substantially slower.

feet. A viewer in Aragon would need to detect a movement of 25 feet from 11.2 miles away, without binoculars — an impossible feat.

It is also possible that Jack meant that the lights moved upward over the course of time during which he and his cousin were viewing them. For example, a viewing episode might begin after sunset when a light would first appear at a location in the general vicinity of the highlands of Cuesta del Burro. In a few minutes, the light would go out. Periodically, in the next hour or two, there would be repeated appearances and extinctions of lights in the same general location, but they would seem to move higher with the passage of time during the evening.

There are explanations for each of these interpretations. The second case, where lights move up slowly over the course of the evening, is the easier of the two possible interpretations to explain. In fact we have already produced that explanation in *Chapter 9 – Larger Than Life*. There we showed that, during the evening, the nocturnal temperature inversion (NTI) develops and produces a refraction of the light passing through it. Recall, we showed that the bending of light rays leads to a condition called "looming" in which light appears to come from a higher position than its actual location. That explanation can also be applied to the Aragon lights.

The other interpretation of what Jack Reed meant is much more difficult to explain. To have lights move up rapidly, to "shoot up" like those in *Chapter 10 – Lights on the One O One Flat,* requires knowledge of the micrometeorological conditions that can exist in the Aragon valley. We begin by learning something about the temperatures along RM2810 and in the Aragon valley region.

Everywhere else in this book we have used an air-temperature model for the NTI in which the temperature varied only in the direction upward from the ground. That served our needs well when the location being considered was out in the middle of an open plain, like in *Chapter 9 – Larger Than Life*. But now, the terrain is deeply cut by a valley, and there are very high mountain peaks nearby. In this setting, we will find that the temperature of the air also varies throughout the valley region and along the slopes, as well as upward into the air.

To get some idea of the kind of temperatures occurring between Aragon and where RM2810 tops out by Cuesta del Burro, we placed temperature measuring equipment at two places along RM2810 shown on Figure 11–1 — locations ***b*** (elevation, 5,617 feet) and ***d*** (elevation, 5,000 feet). Air temperatures were recorded throughout the night of March 10, 2011. These temperatures, plus those recorded by equipment at the Marfa Airport (elevation, 4,835 feet), are shown in Figure 11–3.

Sunset occurred at 7 p.m. CST on March 10th, but you can see from Figure 11–3 that the air temperature was already falling rapidly at all the temperature sensors before the sun set. The air temperature at location ***b*** finally began to outpace the other two and hit a low point just before 8 p.m. The reading indicated that the air at ***b*** was a full eight degrees colder than at ***d*** and six degrees colder than at the Marfa Airport.

This is probably not too surprising because we think of higher elevations as being cooler than lower ones. But, then despite our intuition, the air at ***b*** warmed up by three to four degrees in a matter of 20 minutes, while air at ***d*** and the Marfa Airport continued to cool. Notice that the air at ***b*** exhibited three more episodes of rapid cooling, followed by rapid warming before sunrise the next day. Except for these episodes, though, the air temperature at ***b*** stayed relatively close to 45°F from about 8 p.m. until just after sunrise the next day.

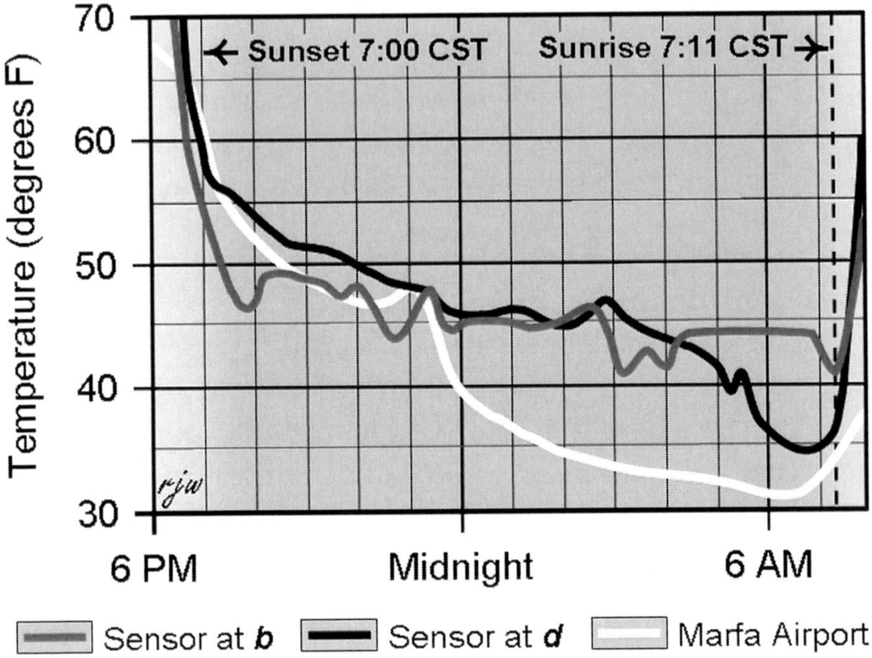

Figure 11–3. Temperatures recorded at Marfa Airport and at locations ***b*** and ***d*** on RM2810 on the night of March 10, 2011.

The rapid movement of the air temperature at ***b*** is indicative of the instability of the air that can occur at other locations along the slopes at night. Those rapidly changing air temperatures could produce the first of the interpretations of Jack's description.

In contrast to the air temperature at ***b***, the air temperature at ***d*** fell steadily until midnight and then stabilized at about the same temperature as at ***b*** until 4 a.m. Starting at 4 a.m., the air temperature then plunged downward by 10°, reaching its low point for the night.

Now notice how differently the air temperature behaved at the Marfa Airport. The decline in airport temperature more or less approximated that occurring at the two sensors on RM2810 until about 11:30 p.m. At that time, it began a 13°–plunge to less than 35°F by 2 a.m. It then remained in the low thirties until the sun's rays reached Marfa in the morning.

The simplest lesson from this experiment is that, in mountainous terrain, the air temperature is a strong function of location. This has major implications for us when considering light projected from high elevations on the slopes of a valley to the valley floor. The light will encounter varying air temperatures along its path. As a consequence, the path the light travels will be bent from a straight line.

But three locations of air temperature measurement do not enable us to predict the light's path. We need a model for the air temperatures in the Aragon region. Fortunately for us, there is information we can draw on.

First, we know from *Chapter 9 – Larger Than Life,* that nighttime air generally increases in temperature as the height above the earth increases. This is the essence of the NTI. It is not an abrupt event occurring over just a few feet as is commonly described in popular literature. In *Larger Than Life,* we presented measured data taken from sensors placed on a 300-foot tower that showed the air temperature changing over the entire 300-foot range.

The other piece of information needed for our air temperature model comes from studies of winds in mountainous terrain. These flows — called *katabatic*[10] winds — commonly occur at night.

The general airflow in an environment of slopes and valley is illustrated in Figure 11–4.

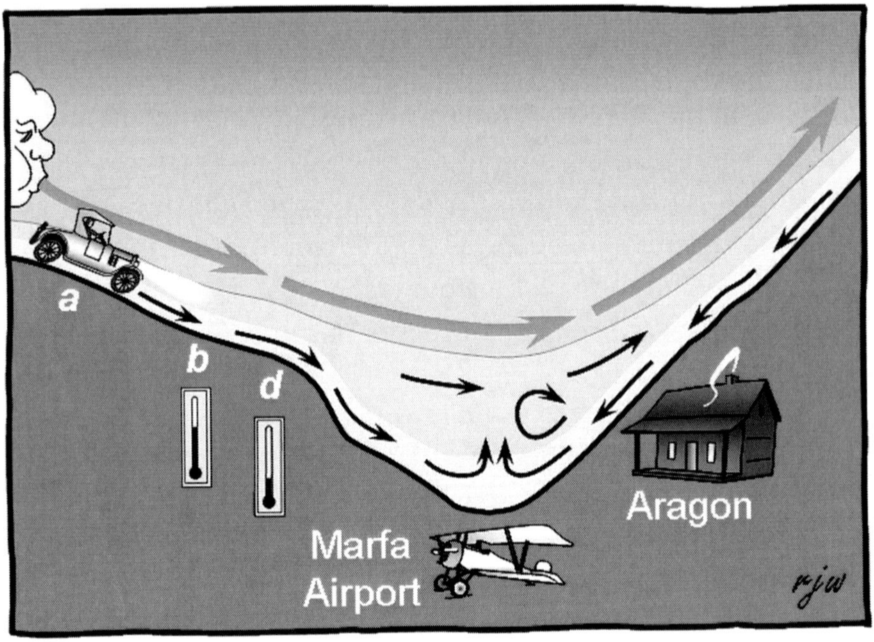

Figure 11–4. Structure of the katabatic winds, large-scale regional winds, and the pooling of cold air in the valley between Cuesta del Burro and the Davis Mountains.

Katabatic winds — stimulated by the cooling of the mountain slopes — flow down the slopes to the valley basin below. As the currents move downward, air along the slopes is cooled by contact with the earth. Once cooled, the air becomes heavier than the adjacent air that is not touching the ground. The heav-

10 *Katabatic* comes from the Greek, where *kata* means downward and *batos* means moving beyond.

ier air then sinks and slides down the mountain slope. These flows appear on both sides of the valley in Figure 11–4 as indicated by arrows in the lighter band of air just above the terrain. The arrows point downhill and go out across the valley before turning upward in the middle.

The same cooling process that is active on the slopes of the valley also goes on in the valley bottom, but the air at the bottom of the valley can't sink farther down so it just stays put if it is not acted on by winds. But what if the elevation of the valley declines as it progresses between the mountains? Then, in addition to winds that flow down the sides of the valley, there will also be winds that flow along the valley floor to the lower elevations. This situation occurs at Aragon. It's downhill to Marfa from Aragon; and in the opposite direction, it's downhill all the way from Aragon to Van Horn.

The thicknesses of the bands of cooler air shown in Figure 11–4 are chosen deliberately. It is known that the streams of cold air adjacent to the mountain slopes can be quite thin — on the order of six to ten feet.[11] In contrast, the depth of cold air in the valley's basin can be substantially greater. There, the depth of cold air increases throughout the night as cold air continues to flow into the basin displacing warmer air to higher positions.

Figure 11–4 also shows the site at Aragon, the Marfa Airport, and locations ***a***, ***b*** and ***d*** on the mountain slope. These five locations do not lie in a single plane. In this figure, we show them in an idealized valley at elevations that represent their relative positions in the cold air.

Knowing how the cold air flows and accumulates, allows us to speculate about the temperatures shown in Figure 11–3.

At the end of the day as the sun moves toward the horizon, the energy it deposits on the earth's surfaces weakens. (Of course, after sunset the energy deposition goes to zero.) Without an influx of energy, the earth's surfaces cool as heat is radiated into space; and in turn, the air contacting the earth is cooled.

Figure 11–3 shows that the air at all three locations cooled at about the same rate until 11:30 p.m. when the Marfa Airport began to cool more than the other two locations. Meteorological data for Marfa on March 10th showed that throughout the afternoon, the winds were out of the south. In the evening, they turned more easterly. But at about 11:30 p.m., the winds turned northerly. The results were dramatic. With the northerly winds, cold air was pushed down from the Davis Mountains, causing the temperature at the Marfa Airport to fall rapidly.

This change in wind direction was ***not*** associated with a weather change. As a general rule at Marfa, no matter what direction the prevailing ***fair-weather*** winds are blowing from in the afternoon, they tend to become northerly near midnight and remain northerly until dawn or mid morning. These are katabatic winds coming down from the Davis Mountains. Marfa is lower than Aragon and locations ***a***, ***b***, and ***d***, so cold air collects there sooner than at the higher locations.

At about 5:30 a.m., the air temperature at ***d*** began to drop rapidly. This was five hours after the rapid cooling at the Marfa Airport. The delayed drop in temperature at location ***d*** was likely brought about by the Aragon valley filling with cold air to a depth of about 200 feet. As the cold air layer became deeper and deeper, it worked its way up the southern slopes of the valley and reached sensor ***d*** about an hour and a half before sunrise. Location ***b*** was 617 feet higher up the slopes and was unaffected by the cold air pooling at the elevations below it.

Katabatic winds are one piece of the puzzle. But to explain how lights appear to move up rapidly, we also need to understand what is happening further up in the atmosphere. In Figure 11–4, higher level winds — illustrated by the large gray arrows — are shown crossing the valley from left to right. These are the southwesterly winds blowing up from Mexico. Depicted here as coming from a blustery cloud, they are the flows of commonly warmer air that is blown across the Rio Grande, over the Chinatis and Cuesta del Burro, across the valley above the cold air inversion layer, and on up over the Davis Mountains. These upper winds moving over the pooled air below, tend to push the top of the pooled air mass to the right in Figure 11–4.

At the same time, the katabatic winds at ground level blowing down both slopes toward the valley below try to push the pooled air rightward on the left side of the valley, and leftward on the right side of the valley. On the right side of the valley, this confluence of katabatic flows and upper southwesterly winds causes somewhat of a rotation of the cold air mass near Aragon. This effect is illustrated by the rotation arrow on the right side of Figure 11–4. On the south side of the valley, there is no large scale rotation because both the upper level winds and the katabatic winds are from the same direction.

The rotation occurring here is not like an eddy, or whirlpool, in a stream of water that goes "round and round". Instead, think of a field of wheat with wind blowing gently over the top. As the wind blows, the wheat bends over. As the wind eases, the wheat straightens back up. The katabatic winds blowing in the opposite direction at the bottom of the cold air mass just increase the effect. The rotations we're talking about are small — perhaps a degree when the winds are right. If the winds were to cease, the rotation would ease back to zero. Or perhaps the rotated air mass would simply pass by and be swept up the slopes of the Davis Mountains.

Knowing how cold air forms and moves in mountainous environments enables us to formulate a model for air temperatures across the Aragon valley and upward from the earth's surface. With a detailed temperature map of the air, we can calculate the path of lights projected toward Aragon and explain Jack Reed's mysterious lights.

The first part of the map is already in place; it is the temperature variation of air upward from the surface of the earth during an NTI. The details of the temperature conditions are those presented in *Chapter 9 – Larger Than Life.*

The other piece of information needed addresses how the air is stratified. Recall, that in our previous applications of the NTI temperature profile, we assumed that the air was rigorously stratified upward from the surface of the earth. In this chapter, we laid the groundwork for changing the stratification assumption when we described rotation of the cold air mass. If the air rotates, the stratification of air layers will also rotate. Instead of the air warming only perpendicular to the earth, the direction of warming will now be at a small angle to the perpendicular.

Because Jack's lights came from the old road, we chose a location on the ***Old Pinto Canyon Road*** from which to project a candidate mysterious light. That location is 17.8 miles from Aragon. The location can be seen in Figure 11–1 as the easternmost tip of section ***b*** on RM2810. At that location, the Old Pinto Canyon Road is about 500 feet west of, and about 25 feet lower than, the current RM2810.[11] Both roads also point toward Aragon.

We then calculated how the light is bent as it traveled through the air to Aragon. Only two factors are included in the refraction of the light — the air temperature at every step along the path, and the angle of stratification at every step. Figure 11–5 shows the results.

The black curve indicates the terrain from the location at the end of ***b*** down to Aragon. The actual terrain that you would encounter by walking the path would be extremely irregular with small, sharp peaks and crevasses. The black curve is a smooth approximation to the "high spots" along the path. But it should be emphasized that the terrain indicated in Figure 11–5 is not just an artistic conception. It is a good representation of the heights one would find on an accurate topographic map.

As the light propagates down the slopes, it encounters air that changes in temperature along its path. As we've said before, the air is coldest next to the earth and warms at heights above the surface. The temperature variation is illustrated by color shading in which the coldest air (lighter band near the earth) fades away into the darkness of the warmer air above. The calculations shown in Figure 11–5 were based on an air temperature near the ground of about 37°F and a temperature 200 feet up in the air flowing from Mexico of nearly 60°F.

If air temperature profile were the only factor refracting the light, we'd find that the light would be bent only moderately. The apparent "looming" of the light on the mountainside would only be about 90 feet. To Jack, that would increase the

[11] The distance between the two roads is about the width of the line representing RM2810 in Figure 11–1.

apparent position of the light from 676 feet to 766 feet. Since Jack and his cousin were 17.8 miles away from the source of the light, it's possible that they wouldn't even have noticed such a small increase.

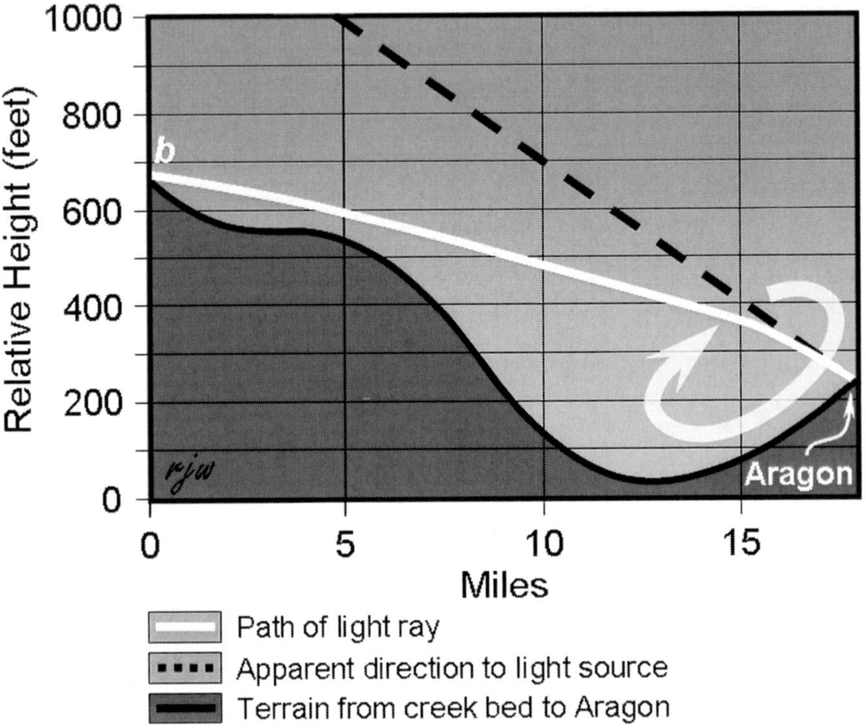

Figure 11–5. Light, shown by the white line, executes an arced path from location ***b*** below Cuesta del Burro, down the mountain slopes to Aragon. Notice how the arcing increases as the light enters the rotating air on the right side of the valley. The black line represents the smoothed surface of the terrain below the light. The dashed black line indicates the apparent direction of the light to a viewer when sighting along the angle at which the light reaches Aragon.

But there is the second effect — the rotated air on the right side of the valley. To simulate that effect, we introduced a rotation of the stratification starting at about the twelfth mile. The rotation increased from zero at mile 12 up to 1.2° at Aragon. The looping arrow in Figure 11–5 indicates the region affected. To picture the effect of this phenomenon, imagine a blob of air 200–300 feet thick and five to eight miles long, that is in place on the slope going north. The blob is executing a slow, organized rotation.

The ray of light projected from the heights of Cuesta del Burro is shown in white in Figure 11–5. In fact there are rays projected in a range of directions from the light source, but the white ray shown is the one that arrives at Jack's eyes on the porch in Aragon.

The ray leaves its source (***b***) at an angle just below horizontal. For the first 12 miles, it bends only a tenth of a degree or so. But as it proceeds onward from there, the angle it makes with the interfaces between layers of air becomes shallower. This condition is key to strong refraction. Remember that in *Chapter 8 – Bending Light Beams*, we showed that as the light ray became tangent to an interface, the ray's path could be bent by tenths of a degree at a single interface. The photograph on the cover of this book is a measure of the effect. The same effect is occurring on the right-hand side of Figure 11–5. Bending of the ray's path is very evident. In this case, the ray is bent about one-half degree down from the initial angle it had when it left the slope of Cuesta del Burro (***b***).

How much the light path is bent by the rotated air mass is highly variable. In order to calculate the exact amount of bending, you would need the specific circumstances of the air on the night in question. Obviously, that would be impossible. Instead, we constructed representative *conditions* and adjusted the *parameters* to get just one-half degree of bending. The impact of the rotated air on the light ray is so strong that less bending or more bending is easily possible.

We chose a half degree of bending because that is the angular width of the moon, and we can use it to visualize the effect. Jack and his cousin couldn't tell that the white light ray's path had bent. What they saw was the angle at which the light was traveling when it reached their eyes. That angle is represented by the dashed black line in Figure 11–5.

If you look at the moon and imagine that Cuesta del Burro — ***b***, where the light originated — is at the bottom of the moon, then the location where Jack perceived the light to be is at the moon's top. When the dashed line passes over the light source on Cuesta del Burro, it is over 600 feet higher than the actual light source. That's a lot of looming!

There is one point we've touched on above, that now needs to be emphasized —

Bending of the light ray's path can go from a small amount to a considerable amount in a matter of minutes.

This is brought about by the rotation of the air mass near the viewer. And, the degree of rotation can change rapidly. For example, one evening, we were filming lights at the MLVC. It was cold, and there were little or no sustained winds. Then, without warning, very cold air began to blow **downward** on us. We don't mean downward from a slope. We mean straight down from overhead. This lasted for only a minute or so, and then was gone. Clearly, the atmosphere was churning somewhat above us.

We mentioned earlier that Jack didn't give any indication of the time period over which the lights moved upward. Consequently, we offered two possibilities — movement over a long period of time and movement over a short period of time.

In the first case, we suggest that the lights moved upward as the NTI grew. Early in the evening, the inversion had little effect. Later, the light bending reached a large value. The virtue of this explanation is that anyone watching the display perceives lights that generally move upward *only*. This situation doesn't require that the lights move downward by an amount comparable to the upward movement during the time that a person is watching the display.

In the second case, we suggest that after the NTI is in place, winds (katabatic and otherwise) modulate the ***direction*** of the stratification in air temperature. Working in concert with one another, these winds produce an organized rotation of a large mass of air. But, they aren't persistent, and so the deflection of light is transient. Because the winds come and go in the short term, and the organized rotations eventually become disorganized, we would expect the light to move both up and down — something Jack didn't mention.

What Jack would have seen if the winds nudged the air to rotate and brought the stratification angle into alignment with the light's path is a ray that steadily moved upward. This might occur over a period of a minute or two. If the rotation became too extreme, or the internal motion of the air mass became disorganized, the ray would snap back down and appear to come from the actual position on the mountain slope.

Perhaps that's why Jack only spoke of seeing lights "go up". He never saw them "go down". After going up, they just reappeared down in their original position until they drove out of view.

Jack Reed and his cousin were correct to suspect headlights. But they couldn't have imagined that the lights they thought were in the Chinatis actually came from autos on the road south of them — nor how unique their story would be in the history of mysterious light sightings on the Marfa highlands.

—§—

11 — Sources

[1] A. Michael Powell, *Trees and Shrubs of the Trans-Pecos and Adjacent Areas.* (Austin, TX: University of Texas Press, 1998), 48.

[2] Cecilia Thompson, *History of Marfa and Presidio County, Texas 1535-1946, Volume 1, 1535-1900* (Marfa, TX: The Presidio County Historical Commission, 1985), 194.

[3] *Southern Pacific Lines, The Galveston, Harrisburg & San Antonio Railway, Brief History of El Paso Division 1881-1925,* compiled by Hugh B. Wilson. (El Paso, TX: [n.p.], 1925), 57.

[4] "Presidio," *The Galveston Daily News*, December 17, 1882, p. 7, col. 5.

[5] "Foreign Intelligence. Railroads Interrupted by Floods." *Galveston Daily News*, November 2, 1879, p.1, col. 6.

[6] Alexander De Laborde, *VIEW OF SPAIN; Comprising a Descriptive Itinerary, of Each Province, and a General Statistical Account of the Country; Including Its Population, Agriculture, Manufactures, Commerce, and Finances; Its Government, Civil, and Ecclesiastical Establishments; The State of the Arts, Sciences, and Literature; Its Manners, Customs, Natural History, etc.* (London: Longman, Hurst, Rees, and Orme, 1809), 83.

[7] Richard Stephen Charnock, *Brad's Illustrated Hand-book to Spain and Portugal: A Complete Guide for Travellers in the Peninsula. With Maps, Town Plans, and Steel Illustrations* (London: W.J. Adams, 1867), 110.

[8] Judith M. Brueske, *The Marfa Lights — Being a collection of first-hand accounts by people who have seen the lights close-up or in*

unusual circumstances, and related material, revised ed. (Alpine, TX: Ocotillo Enterprises, 1989), 23.

[9] "Geological Survey of the El Paso Division (As of April 30, 1925)," in *Southern Pacific Lines, The Galveston, Harrisburg & San Antonio Railway, Brief History of El Paso Division 1881-1925,* 30.

[10] District Traffics Maps 1936–1990 [CD], (Austin, TX: Texas Department of Transportation, 2004).

[11] C. Clements, C. Whiteman, and J. Horel, "Cold-Air-Pool Structure and Evolution in a Mountain Basin: Peter Sinks, Utah," *Journal of Applied Meteorology*, 42 (2002): 752.

12 — Diamond Dust ~ Bearer of Moonbeams

AT ABOUT 10 P.M. ONE NIGHT in May 1987, Marilyn Olsen and her mother, Shirley Sowder, were returning to Marfa from Alpine along US90. The two women had made this drive many times before, and there was nothing special about this particular trip.

After going through Paisano Pass, they rounded the curve on US90 and headed down the grade toward the overpass of the former Kansas City, Mexico, and Orient railroad tracks. After the curve, the highway straightened out and went pretty much westerly.

The next three-mile stretch of US90 is very straight. It runs parallel to the Southern Pacific Railroad tracks, and goes right past the Marfa Lights Viewing Center before turning slightly more northerly. The highway then continues in a straight line for another three miles going past Nopal Road and on to the outskirts of Marfa. (This stretch of US90 is shown on the cover of this book.)

Marilyn Olsen was driving. She looked in her rearview mirror and saw something she had never seen before — something at once amazing and frightening. Large vividly colored balls of light were pursuing their car!

Learning of Marilyn's and Shirley's experience, Judith Brueske interviewed them and published their story in her book, *The Marfa Lights.*[1] What set this sighting apart from all the others we'd read was the women's description of seeing vividly multi-colored lights. Most of the reports of mysterious lights in the Marfa area describe lights that are essentially round and single colored — usually white.

Marilyn was the first to see the lights in her rearview mirror. Her perception was that they were right on top of the back window — perhaps a car length or less behind them. She described the lights as two very bright balls about a foot in diameter, stacked on top of each other. One was "stoplight red" and the other green; Marilyn couldn't recall which color was on top. She estimated that the lights followed their car for three to five minutes.

Hearing her daughter's exclamation, Shirley turned and

saw the display out the rear window. She described the lights as being about 30 feet above the ground. They were the wrong color and too large and bright to be coming from another vehicle. One was red and the other was an "'almost fluorescent' bluish-green" according to Shirley.

What could possibly have appeared as vividly colored balls, one on top of the other? There certainly aren't many natural possibilities. One unnatural explanation was suggested to the women after they reached Marfa — helicopter lights. But the women dismissed that possibility because they hadn't heard anything.

Their dismissal of the helicopter is telling. It suggests that it was too dark that night to see anything outside their vehicle. If they could see anything at all, they surely would have seen a helicopter that was a car length or less behind them and thirty feet up in the air. But, the women chose to refer to the absence of sound rather than the lack of visibility.

As we unravel this mystery, we will see that the darkness figured strongly in what the women saw.[1] It is also key to defining the date when the mysterious lights were seen.

Let's begin by setting the stage for what was going on in southwest Texas — and Marfa in particular — the night of Marilyn's and Shirley's encounter with these mysterious lights.

---

[1] When there is extreme darkness, one must be cautious about interpreting statements regarding height, size, distance, and brightness. In complete darkness with no visible objects in the distance for reference, a one-inch diameter flashlight, that is nine feet away from you, appears to be the same size as the sun (which is 93 million miles away) and the moon (239,000 miles). Their diameters all appear to the viewer to be just over one-half of a degree from side to side.

The night in question we will specify as May 14th. (We will explain why we chose that date later.) National weather maps for May 13–15 from both the National Oceanic and Atmospheric Administration (NOAA) and *The Dallas Morning News* showed a warm front coming from the southwest. The maps indicated no rain in the Big Bend region of Texas on May 13th. But by May 15th, essentially the whole state of Texas was forecast to be covered with a general rain.

The front reaching Marfa turned out to be a weak one. In their hourly log on May 13th, the Marfa Airport recorded clear skies, except for a few clouds at 9 a.m. On May 14th, scattered clouds were reported at 7 and 9 in the morning.

In a monthly report to NOAA, the co-op observation station, Marfa–2, reported a trace of rain at 7 p.m. on May 14th, and thunder was heard in town. More importantly, though, Marfa Airport recorded a wind shift between 6 and 8 p.m., with the winds swinging from nearly southerly to northeasterly. Between 7 and 8 in the evening, the temperature dropped 19°F, and the humidity more than doubled between 7 and 10 p.m. — going from 18% to 40%.

The humidity, blowing in from the east, fueled ice particle development in clouds above. The NOAA report showed that hail was observed at Marfa–2 sometime between 6 p.m. on May 14 and 6 p.m. on May 15. Overhead, in the upper reaches of the troposphere at 30–35 thousand feet above sea level (asl), the winds ambled along at a leisurely 30 mph. At that level, they were more or less westerly throughout the night. Air temperatures at those altitudes were –40° to –56°F (–40° to –49°C).

There were lots of clouds throughout the Trans–Pecos region on the night of May 14. Probabilities of coverage varied by cloud type, location, and the hour during the evening. In general, clouds, rain, and wind chances increased with distance eastward from Marfa toward San Antonio. Precipitation amounts of up to one-half inch were indicated along a 30 to 40 mile-wide band from about Sanderson to Del Rio between 9 p.m. and midnight. But no rain was indicated for Marfa, nor for regions along a line toward the coming moonrise. A line from Marfa toward the impending moonrise ran right along the western edge of the disturbed weather. Twenty to thirty miles variation in the position of the weather front could move probabilities up or down by 50 percentage points along the line toward the moon.

The weather conditions at Marfa were right for the formation of a type of cloud called *cirrus.* Cirrus clouds are the thin, wispy ones[2] that can develop when a warm front is lifted up and over a colder, more stationary air mass.[3] They occur at very high altitudes and appear as long, filmy streaks, oriented along the direction of the air flow. In this case, they would have been oriented with their long direction along a west-to-east line. And, as we shall see, cirrus clouds were the star perform-

---

2 The description given here is strictly applicable only to the cloud type called *cirrus*. But there are three genera of high clouds — *cirrus, cirrostratus*, and *cirrocumulus*. We will loosely refer to any member of these cloud forms as a *cirrus cloud* in this book.

3 Cirrus clouds have long been associated with an impending change in the weather. A very old reference to cirrus clouds — with a great title! — comes from Sir Richard Phillips, *A Million of Facts, of Correct Data, and Elementary Constants, in the Entire Circle of the Sciences, and on all Subjects of Speculation and Practice:* "The Cirrus, or curling cloud, is always uppermost, and often 5 or 6 miles high. It portends rain and wind."[2]

Even older references to the appearance of skies/clouds and their relation to future weather can be found in the *Bible* (Mathew 16:2-3), and in Shakespeare's sonnet, "Venus and Adonis".

ers for this night's drama.

A popular name for cirrus clouds is *mares' tails* because their long flowing appearance is reminiscent of a horse's tail. In fact, the cloud name comes from the Latin word, *cirrus,* meaning *curl of hair*. While cirrus clouds can occur all over the earth, their formation is enhanced in mountainous regions. In those locations, surface air that is blown up and over the mountains can create disturbances in the pressure above, and on the leeside of the mountains. In turn, the pressure disturbances can enhance the formation of water droplets from atmospheric moisture and also promote the subsequent freezing into ice crystals. These ice crystals can be very tiny — comparable in thickness to a piece of aluminum foil food wrap (0.001 inches, or 25 microns).

As the map in Figure 12–1 shows, cirrus clouds are prevalent throughout the western region of the United States to a greater or lesser degree. The highest probability of cirrus clouds occurs where the mountains are highest, *e.g.*, the Rocky Mountains.

Marfa sits squarely inside a gray zone where the probability of daytime cirrus cloud formation is 28%. Surrounding Marfa, a region of high probability extends approximately 100 miles in every direction — making cirrus clouds a common feature in the skies of Alpine and Marfa. On May 14, 1987, the naturally occurring propensity to form cirrus clouds over Marfa was enhanced by the colliding air masses.

At mid-latitudes, such as Marfa (30° north latitude), cirrus clouds may occur as low as about 16,000 feet asl. Or at the other extreme, they can extend upward to 45,000 feet. Temperatures at those high altitudes are very cold, on the order of -90°F (-70°C).

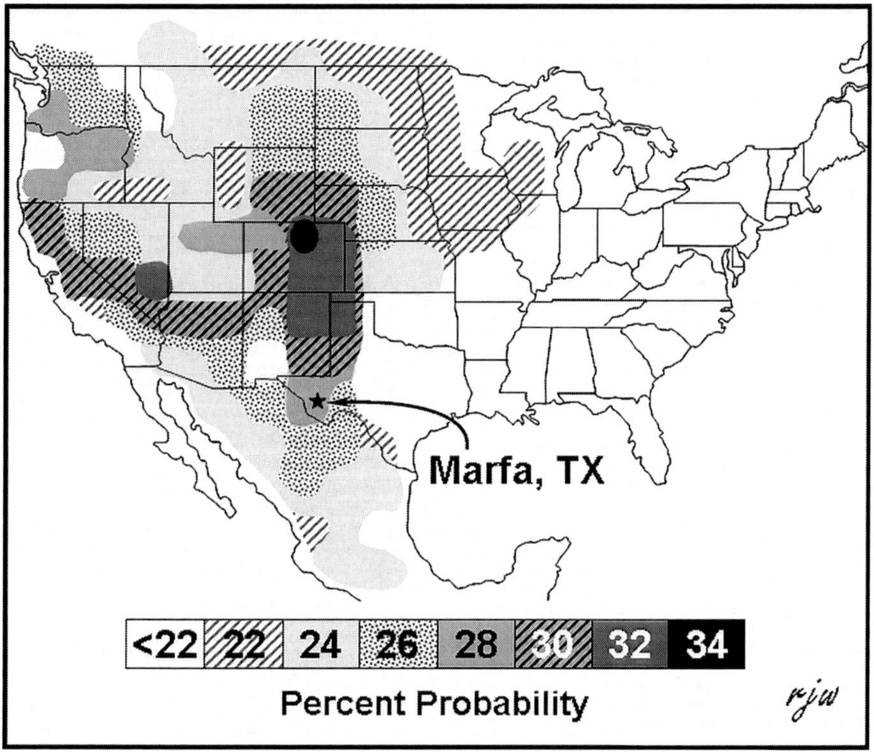

Figure 12–1. Probability of cirrus cloud occurrences during the day for the months of March, April, and May from 1983–2008. The legend shows the percent of the time that cirrus clouds are present at locations across the United States and northern Mexico. (Probability data is from the International Satellite Cloud Climatology Project[3].)

At temperatures between 10°F and –30°F, the clouds are likely to be composed of a mixture of super-cooled water droplets and ice. But below about –30°F, the cirrus clouds are composed of tiny particles of ice. This ice is called ***diamond dust.*** And, cirrus clouds containing diamond dust can produce displays of vivid, iridescent (fluorescent) colors — some of which might be described as having one ball of color on top of another.

The displays can be created in several different ways — (1) when light passes ***through*** an ice crystal and ***refracts***, (2) when light passes ***around*** a crystal and ***diffracts*** (see *Glossary*), or (3) when light ***strikes*** a crystal face and ***reflects***. Although some of these light phenomena can produce displays like those Marilyn and Shirley saw, when we consider all the factors at play on the night of May 14th, we have to eliminate all processes except the one employed in the creation of a phenomenon called a *light pillar* — reflection.

Solar, lunar, and terrestrial light pillars occur commonly. Most are very bright displays. They are produced predominantly by reflection of light from diamond dust particles that take the form of large, flat hexagonal plates. The plates have dimensions from side to side of perhaps 100 microns (0.004 inches). In contrast, the plates have a thickness of only a few microns.

It is light reflecting off the large faces of the crystal plates that is responsible for creating a display called a *pillar*. Light pillars are not a physical edifice in the sky. They are simply the collection of reflections above or below some source of light. Due to the geometric arrangement of light source, reflectors, and viewer, the displays appear as vertical pillars of light.

Because the light is *reflected*, light pillars do not involve any change in the color of the light. The light that a viewer sees coming from a pillar has the same color as the light shining on the diamond dust.

In order to form a pillar, the air must contain ice particles that reflect light without scattering it indiscriminately. Think of a mirror. When the image of an object is reflected by a mirror, we can identify the image as that of the initial object. But if we replace the mirror with a multifaceted piece of cut glass or a diamond ring, we are no longer able to identify the initial image because the facets scatter/reflect the light in many directions. That's good for cut glass and diamond rings, but not good for seeing reflected objects.

The ice particles that create pillars are like a collection of small mirrors. Each particle of diamond dust floats in the air with its largest face being nearly horizontal. As the crystals slowly fall, aerodynamic forces jostle them causing the broad, flat faces of the crystals to wobble up to a few degrees from horizontal. Thus, to within a few degrees, the crystal faces are all pretty much aligned with one another, enabling the ensemble to act like one large, dimpled mirror.

In Marfa, Texas, you only need to go a few miles up in the atmosphere to find the temperatures needed to make diamond dust. When the atmospheric conditions are right up there, you can find millions and millions of hexagonally-shaped platelets of diamond dust. In Figure 12–2, we've singled out one flake to illustrate how an individual crystal of diamond dust contributes to the formation of a luminous display seen from the ground.

The woman on the right is looking toward the diamond dust hexagon. On the left, over the horizon and out of her sight, the moon is bathing the earth in light. Its rays are intercepting all the diamond dust crystals in the sky above. And in this case, the white light ray depicted by the arrow is illuminating our hexagonal crystal.

Figure 12–2. Illustration of the path followed by a specific ray of moonlight from a spot on the moon that is reflected from a flat, hexagonal ice crystal floating in the air.[4]

This ray of light comes from a specific location on the surface of the moon and strikes our diamond dust crystal on its

[4] Bear in mind that Figure 12–2 is a highly distorted cartoon intended to illustrate the principals. It does not accurately portray the relative positions of the elements to one another. For example, the clouds containing the diamond dust could cover a distance of 100 miles but only be a few miles above the earth. Additionally, while the moon seems to be just over the horizon from the woman, in reality, it is over 200,000 miles out in space and appears to us to be only a half degree wide. Angular relations are misleading. In the illustration, our observer is shown looking up at the hexagonal ice crystal but actually she would be looking almost horizontally.

broad, flat face which is shown (at this instant) **oriented in a fixed attitude** relative to the earth. The crystal's surface reflects the light downward at the same angle that the moonlight hit the crystal. Because of the orientation of the diamond-dust crystal, the center of the reflected light beam shines on the woman, and she sees a glint of moonlight.

But the description above is static. This single instance of alignment of moon spot, diamond dust, and observer is altered by the aerodynamic forces that cause diamond dust crystals to wobble about their axes as they fall. In Figure 12–2, the axis is illustrated by a solid line extending upward from the top face of the crystal. The two dashed lines on either side of the axis show the extremes of a cone inside of which the axis wobbles. This wobble is essential to the creation of an image for the viewer. As the crystal wobbles, it continuously sweeps out new areas on the moon that are in the right locations to line up with the viewing direction of the woman.

There's not much light to be seen when the spot shines on the observer. After all, the crystal is tiny, and it's tens of miles away. But there are millions and millions of these crystals wobbling around and shining toward the observer bringing rays of light — not just from one spot on the moon — but from all over the moon. Collectively, they produce a bright light as they shine toward the observer from **all along** the diamond dust cloud.

But what the woman sees is not a *picture* of the moon. It is the *light* of the moon. As each single plate of diamond dust wobbles about, it projects moonlight from whatever spot on the moon happens to be aligned with the viewing direction of the woman. She sees moonlight — but not light organized as an image of the moon.

The arrangement of moon, diamond dust, and observer illustrated in Figure 12–2 is essential to the creation of lunar pillars and dictates the time of day in which they occur. Since the diamond dust is restricted to an ***attitude*** in the air that is essentially horizontal, the rest of the geometry involving light rays and reflection requires the moon to be on a line near the horizon — that is, either rising or setting. Indeed, if you search the Internet for photographs of sun and moon pillars, you'll find that all of them occurred within a few degrees of the sun or the moon crossing the horizon.

Sun and moon pillars are usually white. That's because the light shining on the ice crystals is usually white. Pillars simply reflect the color of whatever light shines on the diamond dust. In the case of Marilyn and Shirley, the light needs to be "stop-light red" in one "ball" and an "almost fluorescent bluish-green" in the other.

So how did Mother Nature create that collection of colors in the display?

The red light reaching Marilyn and Shirley came about the same way that red sunsets and sunrises occur all the time — particles of dust low in the atmosphere scattered out the blue and green colors. But the scattering on the evening of May 14, 1987, may have been enhanced over the normal reddening of the evening skies by a dust storm occurring along the Mexican side of the Rio Grande. The dust storm was so strong that on NOAA's weather reanalysis web page,[4] they indicated reduced visibility at 6 p.m. in northern Coahuila, Mexico. At its center, visibility was reduced by about 30%, and some degree of reduction was indicated over an area 130-miles in diameter.

The location of the storm was nearly on a line from Marfa

toward the direction where the moon would rise later that night. Though the wind storm was gone by the time Marilyn and Shirley made their drive to Marfa, some dust from the storm would have remained in the atmosphere. Near the earth, this residual dust would have enhanced the reddening of any white light shining through it. This effect is commonplace in the desert southwest.

The "bluish-green" light that the ladies saw was also likely created by the dust storm, but from very uncommon circumstances. Precisely how it occurred is impossible to determine because there are no records of many of the controlling factors. But we know some things that can be used to illustrate the general sequence.

Imagine that a substantial amount of dust was lofted high into the air (20–30 thousand feet). We know from storms in the Gobi Desert of China that dust can reach extreme heights. In fact, dust from Gobi storms has been lofted high enough in the atmosphere to be blown all the way across the Pacific Ocean to the United States.

Larger dust particles (say one micron diameter) from the Coahuila dust storm would preferentially scatter red light. If they were lofted up to the 20–30 thousand foot level, they could directly produce the blue-green color of the moonlight passing through at that altitude.

Sound improbable?

Well, let's consider some specific examples where particulates in the atmosphere altered the color of the moon and sun light. In the first case, smoke from forest fires burning in Alberta, Canada on September 23, 1950 was carried all the way east to Europe.[5] On the afternoon of September 26, 1950, the smoke cloud — which was between 31,000 and 43,000 feet asl — was west of Edinburg, Scotland. An astronomer at an observatory there noticed that the sun's light had become "a deep indigo blue" and rushed to prepare his telescope so he could measure the spectrum of the light he was seeing.

When he compared the spectrum he'd measured to that of normal sunlight, he observed that the red portion of the sunlight had been suppressed relative to the blue and green components. His analysis led him to suggest that the clouds contained particles of oil — cooked out of the trees in the Canadian fire — which were about a micron in diameter. That diameter is close to twice the wavelength of red light and should preferentially scatter red light.

What's so special about this occurrence was that an astronomer saw the spectacle and could measure the spectrum. When he published his findings a year later,[5] he reported that the blue sun had also been seen by people in the United States and Canada. What's more, the phenomenon continued into the night at his location where he also witnessed a blue moon. That must have been very impressive since, coincidentally; it was the night of a full moon with 100% of the moon's disc illuminated. The next day, the clouds had passed on eastward, and the sun and moon returned to their normal appearance.

There are many, many examples of people seeing blue or green suns and moons. The one above was suggested to be due to oil drops. But the phenomenon can be caused by water drops. Consider the story of a Mr. D. Pidgeon in London, who was looking out his window at the sun one afternoon in February, 1888, when a train went past and filled the air with exhaust steam. While the steam was present, the sun shone through it with a "vivid light-green colour".[6] (The ability of steam to produce a green appearing sun was already well known by 1884.[7])

In these two cases, it seems likely that the coloring of the sunlight was caused by oil drops and water drops. But dust was the cause of the most widely reported observations of colored sun sightings in recorded history. The event was the eruption of the volcano, Krakatoa, in Indonesia on August 26–27,1883.[5] Dust ejected into the air caused widespread appearances of a blue or green sun and moon around the world. There was so much dust, and it was ejected so high, that 26 days later it was still present, having circled the globe twice!

Now that we know the phenomenon can exist, let's apply the facts about moon colors and reflecting properties of diamond dust to the sighting reported by Marilyn and Shirley.

Neither Marilyn Olsen nor Shirley Sowder could remember what day in May their observation took place. However, it most likely was May 14th. Not only were the weather conditions right for the formation of cirrus and other clouds, but the sun had set at 8:42 p.m. CDT, and the moon did not rise until 10:35 p.m. — making it very dark during the women's drive.

At 10 p.m., the moon was approximately seven degrees below the eastern horizon. At 10:33 p.m., it was still about one degree below the eastern horizon and had a bearing from due north of approximately 122°. This bearing should be contrasted to the 104°-bearing of the straight stretch of US90 that

---

[5] In 1888, the Royal Society published a 500-page analysis of the phenomena associated with the eruption.[8] Three hundred and twelve pages were devoted to the extraordinary sightings of a blue or green sun and moon.

runs past Nopal Road and on into Marfa. Amazingly, this places the moon below the horizon and almost directly behind Marilyn and Shirley! Perfect for the display that is about to unfold in the clouds above them.

Figure 12–3 below and on the back cover illustrates the likely scenario. There is no moonlight shining on the earth where the ladies' car is. In fact, the moon is off the far lower, left corner of the figure, well below the horizon. Its rays (represented by the arrows labeled *green* and *red*) are only "visible" to the clouds.

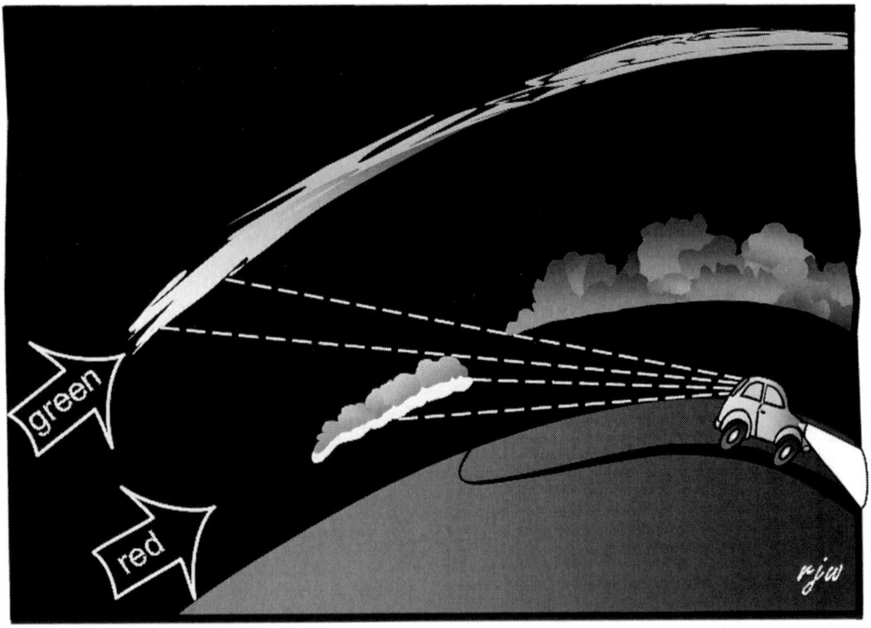

Figure 12–3. Configuration of moonlight, layered clouds, and Marilyn and Shirley's vehicle at about 10:30 p.m. CDT on May 14, 1987. Behind the car, the bearing of US 90 is 104°, and the moonlight is coming from a direction with a bearing of ~122°. (See the back cover for a color rendition of this figure.)

The clouds that night occurred at low, medium, and high altitudes, and their probabilities of formation varied with location.[4] As illustrated on the right side of Figure 12–3, starting at about 12,000 feet, thick medium elevation clouds have formed over Marfa beginning around 9 p.m. This mass of clouds is about 80 miles across and extends all the way to the Chinati Mountains and east-southeast for a distance of about 50 mile before clearing. Cloud coverage throughout the circular region of the medium height clouds is 90% or more. But along the line toward the coming moonrise, the probability of medium height clouds diminishes steadily (going from 90% over Marfa down to ~40% just across the Rio Grande).

To the east 50 to 100 miles away and about 4,000 feet above Marfa's elevation, low stratus clouds have formed as shown in the center of Figure 12–3. These low clouds only have a probability of ~20% over Marfa but their probability of occurrence increases to ~50% near the Rio Grande about 90 miles from Marfa.

Finally, above it all, at 25,000 feet, a canopy of cirrus clouds has formed, covering the entire region from Marfa down across the Mexican border. Their occurrence ranges in probability from 50 to 70% in the area illustrated in Figure 12–3.

On the ground where Marilyn and Shirley are located, it is pitch black. The clouds block out starlight, and the earth blocks direct light from the moon. But up at the altitude of the cirrus clouds, it is very bright indeed. The moon is at 96% of a full moon.

In Figure 12–3, the moonlight in the upper green ray reaching the underside of the cirrus clouds, at 25,000 feet, does not have to propagate through the thick atmosphere close to the earth. That ray is assumed to be coming from one-half degree below the horizon at Marfa. At that angle, the density of the atmosphere is 40% or less of the density at sea level. Consequently, the normal atmospheric scattering of the short wavelengths of light (blue and green) out of the beam of moonlight is minimized.

While the normal atmospheric scattering of blue and green is diminished, the ray experiences enhanced scattering from dust recently lofted into the upper atmosphere. As we described above, this dust removes the red components of the moonlight leaving a blue-green light to shine on the cirrus cloud. The blue-green light then strikes the millions of crystals of diamond dust and reflects down to Marilyn and Shirley.

At the lower left in Figure 12–3 is another ray of light from the moon. In this case, it is colored red from having traveled through the low-level dust left behind by the dust storm at 6 p.m. that evening, plus the dust from continuing winds in Mexico. This ray lights up the underside of the low-level stratus clouds similar to what you would see in a deep red sunset.

Now, think about how this scenario would appear to Marilyn and Shirley looking backward at an angle of about 18° south of US90[6] and witnessing these vivid lights. The green and red clouds in Figure 12–3 are "framed" by surrounding clouds so that they appear as two balls of light — blue-green on top and red on the bottom.

During the moon's final minutes before rising, its ascent

---

6 That angle is slightly toward the driver's side of the vehicle and would be a comfortable viewing angle for either Shirley to see the lights in the rearview mirror or her mother to see them out the rear window of the car.

proceeded at about two-tenths of a degree per minute. Thus, during the three to five minutes that Marilyn estimated they saw the light, the moon would only have risen about one degree. It could have been hidden[7] by the earth during the entirety of Marilyn's and Shirley's observations, and they might have driven on in to Marfa and never even have been aware that moonrise was impending— *and,* never suspecting the moon was responsible for the balls of light.

So incredible was Marilyn's and Shirley's story, that when we spoke to Judith Brueske in July of 2009, she voiced some skepticism about its plausibility, and said she had removed this story from subsequent editions of her book. Over the years that she'd been interviewing people about mysterious lights, no one else had ever described seeing blue/green lights. When someone described colored lights to her, they invariably turned out to be the red and green lights of railroad signals.

But it was their description of the light's fluorescence which made us suspect that the two women did indeed witness a rare and mysterious light phenomenon. They happened to be in the right place on US90, at the right time of night just before moonrise, when meteorological conditions were ripe for the creation of cirrus clouds with the right sized diamond dust.

Think about it. The balls of light Marilyn and Shirley saw that night began their eight-minute journey 93 million miles away on the sun. After reflecting off the moon, the light ran a gauntlet of dust-particle collisions before encountering both,

---

[7] The angle up from the horizontal to the tops of the mountains east of Marfa varies with the position along US90 and the bearing of view. At 122° bearing, the visible horizon angle, as seen from US90 on the two straight stretches of highway described at the start of this chapter, varies from ~0.4° to ~1.0°.

low clouds, and ice crystals in cirrus clouds high up in our atmosphere. The diamond dust reflected the moonlight down to US90 and the rear window of the women's car — creating a startling display. What an incredible trip!

— Addendum —

IN OUR DESCRIPTION OF WHAT MARILYN AND SHIRLEY SAW, we illustrated how the color spectacle could have occurred. Some might question the ability of the moon to serve as a light source for a colored display seen on the earth. There are readily available sites on the Internet which contend that the amount of color in moonlight is too weak to be visible to the human eye. These sources assert that images produced by moonlight can only be perceived as gray tones.

In fact, it is simple to prove that the moon has perceivable color. Just look at the photos labeled *The Moon* and *The Sun* on the back cover. We took both photographs using a prism of water[8] in the path between the camera and the light source. The

8 A water prism was used because water has essentially the same optical index as diamond dust.

spectra displayed look remarkably alike.

Not only are the two spectra capable of being detected with a digital camera, they are also visible to the unaided eye. We looked through the prism directly — without any camera in the path (but with an additional dark filter in the case of the sun) — and we saw the same spectra.[9] Clearly moonlight enjoys the same range of spectral components as the sun.

—§—

12 — Sources

[1] Judith M. Brueske, *The Marfa Lights — Being a collection of first-hand accounts by people who have seen the lights close-up or in unusual circumstances, and related material,* revised ed. (Alpine, TX: Ocotillo Enterprises, 1989), 15.

[2] Sir Richard Phillips, *A Million of Facts, of Correct Data, and Elementary Constants, in the Entire Circle of the Sciences, and on all Subjects of Speculation and Practice,* (London: Darton & Co., 1848), column 543.

[3] NASA, "International Satellite Cloud Climatology Project (ISCCP),"
http://isccp.giss.nasa.gov/

[4] NOAA Earth System Research Laboratory, Physical Sciences Division, "3-Hourly NCEP North American Regional Reanalysis (NARR) Composites," NOAA,
http://www.esrl.noaa.gov/psd//cgi-bin/data/narr/plothour.pl

---

9 Readers can perform the same experiment with a common glass prism. Looking directly at the moon through the prism, you will see the spectrum of colors. In the case of the sun, the rays should be projected onto another surface, such as a sidewalk or a wall. **NEVER look directly at the sun!**

[5] R. Wilson, "The Blue Sun of 1950 September," *Monthly Notices of the Royal Astronomical Society,* 111, (1951): 477.

[6] D. Pidgeon, "A Green Sun," *Nature*, 37, March 8, 1888, 440.

[7] C. Michie Smith, "Observations on a Green Sun, and Associated Phenomena," *Nature,* 30, August 7, 1884, 347. (Abstract of a paper read before the Royal Society of Edinburg, July 7, 1884.)

[8] G. J. Symons, F.R.S., editor, "The Eruption of Krakatoa, and Subsequent Phenomena," *Report of the Krakatoa Committee of the Royal Society* (Ludgate Hill, London: Trubner and Company, 1988), 57 and 59.

13 — Conclusion

WHEN WE BEGAN THIS INVESTIGATION, we hoped to find a brand-new source of light. Knowing that the odds of finding such a light were distinctly against us, we hoped at least to discover that a known, but obscure source was responsible for the mysterious lights. Alas, neither hope was realized.

Instead, we found that generically the physical process responsible for all the lights was the thermal agitation of the molecules found in very hot matter. By *hot,* we mean matter with temperatures in the thousands of degrees. All matter with such elevated temperatures emits light. For some of the lights examined, the ultimate source was the sun. For all the others, it was filaments of the metal — tungsten. These sources of light are well known. But the linkages between them and the lights on the Marfa plain are not.

Consider the sun. At a surface temperature of about 5,500°C, it emits a broad spectrum of light. That light travels out into the universe. In our direction, it travels 93 million miles, becoming weaker as it advances. The portion of sunshine that interests us either shines directly on the earth or takes a side trip to the moon. Of the light which hits the moon, about ten percent is reflected from the moon's surface sending a full spectrum of light onward another 230,000 miles to the earth.

In this book, sunshine was the ultimate source of mysterious lights in the mirages described in *Chapter 8 – Bending Light Beams*, the looming behavior described in *Chapter 9 – Larger Than Life*, and the colored balls seen by Marilyn and Shirley in *Chapter 12 – Diamond Dust ~ Bearer of Moonbeams*. As mysterious and "unnatural" as these effects appeared to the observers, they all derived from purely natural phenomena.

In contrast, the lights most commonly seen on the Marfa plain are the result of a phenomenal technological accomplishment — the tungsten filament. What began with the harnessing of electricity in the late 19th century was enhanced in the early 20th century by the discovery that when electricity passed through a tungsten filament, heating it, light was emitted in a controlled manner. Encapsulation of the filament in an evacuated glass bulb rendered the light source into a usable form. The high utility of controlled light sources quickly led to commercialization of bulbs and the generation of electricity. Electric lighting became ubiquitous. Incorporation of these lights into automobiles in the 1910s now made high-intensity lights mobile. And, that was the last thing needed to provide mysterious sightings.

But mysterious lights don't bloom everywhere there is

sunshine, moonlight, or an encapsulated tungsten filament. Very special circumstances are required to produce lights that can truly be called *mysterious.* You must have great enough distances between the source of light and the viewer so that the viewer cannot identify the actual source. You must have a highly bright light source with enough intensity to project its rays across great distances. And, you must have an environment and terrain with the right meteorological and climatic conditions to enable refraction and reflection to interact with the light teasing out unique displays.

There are select places on earth that satisfy these requirements. The Marfa plain is one. In the thin, dry air of the Marfa plain, refraction and reflection make the perceived lights do unexplained things, like move upward or appear as large, iridescent balls. Add decades of observation by thousands of people, and a legend of inestimable proportions emerges — the Marfa Lights.

—§—

Appendix

Terminology and Definitions Associated with Light and its Properties

IN *CHAPTER 5 – ILLUMINATING THE MYSTERY*, we gave short definitions of terms associated with light and its properties. This appendix provides an extended discussion of those terms for readers who want more technical information.

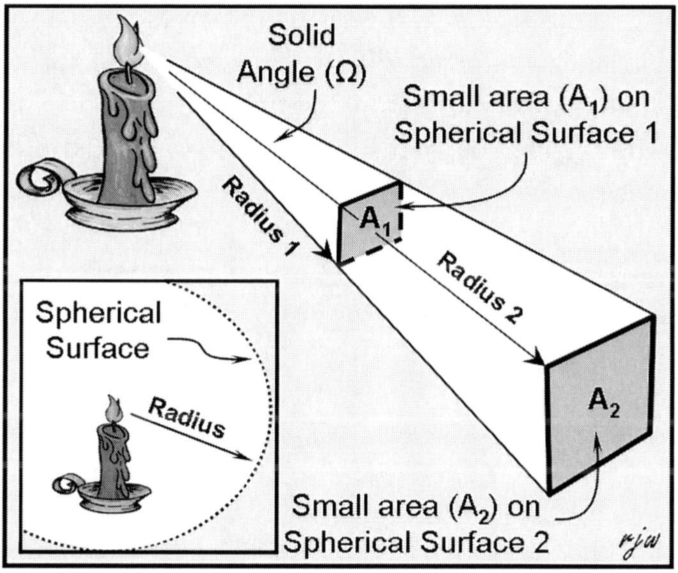

Figure A–1. Important measures of light.

The inset in the lower left of the Figure A–1 shows a small candle with a portion of a white sphere enclosing it. The candle is shown emitting light along a line labeled, *Radius*, toward a curved line labeled, *Spherical Surface*. The spherical surface completely encircles the candle. Light from the small candle flame travels outward in all directions and intercepts the spherical surface surrounding it.

An enlargement of the candle and spherical surface are shown above and to the right of the inset in Figure A–1. In the enlargement three rays (out of four) are shown projecting outward from the flame and intersecting the corners of two areas highlighted in dark gray. About half way out along the rays, a small area (A_1) is designated. Think of this area as a portion of a spherical surface (surface 1) enclosing the candle flame, just as in the inset at the lower left. The distance from the flame out to this surface is labeled *Radius 1* (R_1), and it is the radius of the inner sphere. Farther on along the rays is a larger area (A_2) associated with a larger spherical surface (surface 2) that also encloses the flame. The distance from the flame to the second surface is Radius 2 (R_2) — the radius of the outer sphere.

The tapered white-colored volume defines something called a ***solid angle***. Mathematically, the solid angle can be obtained by dividing one of the areas by the square of the radius to the area. Thus, the solid angle (Ω) is given by —

$$\Omega = A_1/R_1^2 = A_2/R_2^2$$

Notice that the solid angle is the same whether it is calculated from the first area and radius or the second area and radius. The unit of solid angle is called ***steradian***.

Light emitted by the candle flame is a form of power. The power radiated is measured in ***lumens***. Lumens are precisely defined for scientific purposes. But here, we simply need to know that the word *lumen* describes how much power is being emitted in the form of light.

Light from the flame is emitted in all directions. But consider just the rays of light (measured in lumens) emitted into the white tapered region to the right of the flame. If we measure the lumens (L) entering the taper and divide that number by the size of the solid angle (Ω), we obtain a new measure of light — called ***intensity***.

$$I = L/\Omega$$

The light's intensity is measured in lumens per steradian. Intensity has a name of its own — ***candela***. The more lumens (L) that are projected into the solid angle (Ω), the higher the intensity. Also, for a given number of lumens, the smaller the solid angle, the higher the intensity.

Notice, that the intensity is not a function of distance from the flame. Light projected into the solid angle stays in the solid angle, no matter how far it travels to the right in Figure A–1. The light travels in straight lines within the lines of radii that define the solid angle. Just as the solid angle was independent of radius from the flame, intensity is independent of radius from the flame.

The last term we need to define is the ***illuminance*** of light. Consider again the small areas (A_1 and A_2) shown on the spherical surfaces 1 and 2. Illuminance at either of these areas is found by dividing the lumens (L) falling on an area by the size of the area. Thus, illuminances E_1 and E_2, at radii R_1 and R_2, respectively are given by —

$$E_1 = L/A_1$$

and

$$E_2 = L/A_2.$$

Illuminance (E) has the units of lumens per square meter. One lumen per square meter is defined to be one ***lux***.

As we said above, the total lumens (L) projected into the solid angle (Ω) in the figure do not change as the distance from the flame increases. But the size of the area intercepted gets larger as the radius increases. This is evident just by looking at the figure, and is also reflected in the two equations above which show a different area in the denominators.

It is clear from the two equations above that if the total lumens (L) are the same at each radius, but the area A_2 is larger than A_1, then the illuminance E_2 will be smaller than illuminance E_1. In fact, in the ideal case of a very small flame and a large distance from the flame, doubling the propagation distance reduces the illuminance by a factor four. That means the illuminance diminishes as the reciprocal of the square of the propagation distance ($1/R^2$).

The three light terms *lumens, intensity*, and *illuminance* will be used repeatedly in *Chapter 5 – Illuminating the Mystery*. This appendix was provided for the technically interested readers. But the discussions in *Illuminating the Mystery* can be sufficiently understood using only the shorthand definitions below —

Lumens — Light emitted by a source without regard for where the light is projected.

Intensity — Lumens projected into a solid angle, divided by the size of the solid angle. **(candelas)**

Illuminance — Quantity of light falling on a surface perpendicular to the ray, divided by the area of the surface. **(lux)**

Solid angle — An area on the surface of a sphere divided by the square of the radius to the surface of the sphere. **(steradian)**

—§—

Glossary

aerosol: a system of liquid or solid particles, uniformly distributed in a finely divided state through a gas, such as air

arroyo: a gully, draw, or wash through which water may occasionally flow

asl: above sea level

BBTC: Big Bend Telephone Company; owners of the tower seen from the Marfa Lights Viewing Center.

bearing: a term indicating the direction to something from a specific location, such as where a person is standing. The direction is referenced to something that serves as a standard. In this book, two types of standards are mentioned: *true north* and *magnetic north. True north* is the direction to the axis of rotation of the earth. *Magnetic north* is the direction that a compass points. While true north always points toward the axis of rotation of the earth, magnetic north is quite variable across the earth. A true bearing might take the form *97°–true*. That would mean 97° clockwise from the direction of the rotation axis of the earth about the person's location. A magnetic bearing might read *97°–magnetic,* and it would mean 97° clockwise from the direction a compass points **at a person's location**. Generally the two bearings do not point in the same direction.

Big Bend Region: Writing in his diary on March 12, 1849, Lt. William Whiting was the first person to use the term *Big Bend*[1] to describe the change in the direction of the Rio Bravo as it angles from a southeastern course to a northeastern one. Over time the term has expanded to apply to a region, not just a river course. John Gregg in his Master of Arts thesis[2] defined the region – "The river here . . . appears as a wide graceful curve like a half moon with the points up. The land on the Texas side, in the hollow of this curve, is the 'Big Bend'." He went on to say that the region encompasses the Davis, Chinati, and Chisos mountain ranges. The region he described in 1933, which we call the *Big Bend* today, is approximately that contained in the counties of

Jeff Davis, Presidio, and Brewster.

[1] W.H.C. Whiting, "Whiting Diary", *Publications of the Southern History Association* 10, no. 1 (January, 1906): p. 6.

[2] John Ernest Gregg, *The History of Presidio County* [Master of Arts Thesis]. Austin, TX: University of Texas; August 1933. Excerpts reprinted in *Voice of the Mexican Border*, Centennial Edition 8, (1936): p. 11.

caldera: a large crater left behind after a magma chamber below a volcano empties and the overlaying strata collapse into the chamber.

candela: see *Appendix*

candlepower: a light standard, now obsolete, used until the early 1900s to quantify light output. In the middle 1800s, efforts to develop standards for quantifying light were impelled forward by the need to test the quality of gas being sold to residents of London. To this end, the Metropolis Gas Act of 1860 was enacted. In it, a standard candle was specified by the dimensions and composition of its body, wick, and flame. The light output of a candle meeting the specifications would be compared to the light emitted by a flame of gas. A properly adjusted gas flame had to produce the correct candlepower, that is, a specified number of "candles" of light. Other countries in Europe similarly specified their own light standard. None of the "standards" was particularly repeatable, and they weren't equal to one another. Over the decades, efforts to produce a unified standard for "candlepower" led to the definition of the International Candle in 1909. While this agreement specified the relation of one national "standard" to that of another country, it did not produce a primary standard. In 1948, a primary standard was adopted that traced its light to that produced by one square centimeter of a black body held at the freezing point of molten platinum, 1773°C. The amount of light emitted by this standard was **measured** to be the same as that output by 58.9 International Candles. With the introduction of the new standard, a new unit of intensity was also defined. The light output by the black body was **defined** to be 60 New Candles. And the new intensity would be called a *candela*. Thus, one candela equals 58.9/60 International Candles. Or, in today's terminology, 1 candela = 0.981 candlepower.

CDT: central daylight time

charco: a waterhole or pool of water

cirrocumulus clouds: a collection of circus clouds that forms little tufts rather than a continuous sheet. In this format, the clouds contain super-cooled water droplets in addition to ice particles.

cirrostratus clouds: a collection of cirrus clouds that has grown laterally and connected to form a sheet of cirrus. This sheet is called *cirrostratus.*

cirrus clouds: thin wispy clouds that occur at very high altitudes (16–45 thousand feet). The long, filmy streaks of clouds are only a few kilometers thick and their long directions are oriented along the direction of air flow. A popular name for cirrus clouds is *mares' tails* because their long flowing appearance is reminiscent of a horse's tail. The cloud name comes from the Latin word, *cirrus,* meaning *curl of hair*. Cirrus clouds generally have a white to grayish appearance. The clouds are composed of ice crystals. Sometimes the crystals appear in the form of hexagonal structures. When the faces of the crystals take on the regular geometric pattern of a hexagon, they can produce color displays through refraction of light passing through the crystals. The ice crystals of cirrus clouds can be very tiny — comparable in thickness to a piece of aluminum foil food wrap (0.001 inches, or 25 microns).

civil twilight: the period of the day just before sunrise or just after sunset. Morning civil twilight begins when the geometric center of the ***sun*** is 6° below the ***horizon***. It ends at sunrise. Evening civil twilight begins at sunset and ends when the geometric center of the sun reaches 6° below the horizon. The specific time duration of civil twilight is a function of latitude and time of year.

cloud: a visible aggregation of fine water droplets, or tiny ice particles, or a mixture of both. Clouds almost always appear in the lowest layer of the earth's atmosphere, the troposphere.

CML: classic Marfa lights. See *Chapter 6 – "Classic" Marfa Lights*

CST: central standard time

cuesta: a long, low ridge with a relatively steep face or escarpment on one side and a long, gentle slope on the other.

diamond dust: very small particles of ice that fall slowly under the combined influences of gravity and the up/down flow of air. The particles may be in a high cloud, like cirrus. They may also appear in a cloudless portion of the sky. Or, they may exist as a "fog" of ice particles near the earth's surface. In all cases, they are formed in extremely cold air, *e.g.,* –30°F or colder. Certain forms of diamond dust can produce spectacular displays of light in the sky. Two commonly mentioned, but rarely seen, forms of display are the *circumhorizonal arc* and the *circumzenithal arc.* For a picture of a circumhorizonal arc, see— National Geographic, National Geographic News, Thursday, October 28, 2010, "Photo in the News: Rare 'Rainbow' Spotted Over Idaho," http://news.nationalgeographic.com/news/2006/06/060619-rainbow-fire.html.

diffraction: the term applied to the process of a wave encountering an obstacle and moving beyond it. For a small obstacle, the wave could move around it. For an aperture, the wave could pass through it. During the process, portions of the wave may travel different paths to arrive at the same location. This leads to varying degrees of constructive or destructive interference between the sub-waves. The pattern of interference is the manifestation of diffraction.

EST: eastern standard time

flat: a USGS feature classification defined as a relatively level area within a region of greater relief. See USGS Feature class definitions on
http://www.usgs.gov/pubprod/maps.html

illuminance: see *Appendix*

index of refraction: the ratio of the speed of light in vacuum to the speed of light in a medium under consideration. Thus, the index of refraction in vacuum is one. For all optically transmissive materials, the index of refraction is greater than one.

intensity: see *Appendix*

IOR: see *index of refraction*

katabatic: a term applied to "drainage" winds. These are winds produced by higher density air that is pulled downhill by the force of gravity. At night cooler, heavier air, adjacent to the earth, slides down the slopes of hills/mountains producing katabatic winds.

lumen: see *Appendix*

lux: see *Appendix*

MAAF: Marfa Army Air Field — a World War II airfield near the Marfa Lights Viewing Center

magnetic declination: the difference between the direction that the north end of a compass needle points at a specific location and the direction of true north. A positive magnetic declination indicates that the compass needle points east of true north. For example, at Aragon, Texas, the magnetic declination is +7.6°. At Antelope Springs on the Marfa plain, the magnetic declination is +7.5°. To convert from a true bearing to a compass bearing, subtract the local magnetic declination from the true bearing. For example, at Aragon, a true bearing of due south, +180°, would have a compass bearing of +172.4°.

magnetic north: see *bearing.*

micron: a distance of one millionth of a meter (1/1,000,000 meter)

Minimum Angle: see *Snell's Law*

MLVC: Marfa Lights Viewing Center

MVL: mercury vapor lamp

nanometer: a distance of one billionth of a meter (1/1,000,000,000 meter)

NASA: National Aeronautics and Space Administration

NIST: United States National Institute of Standards and Technology

NOAA: National Oceanic and Atmospheric Administration

nocturnal temperature inversion: a condition produced by the overnight radiative cooling of the earth's surface in which the temperature of the atmosphere increases with altitude in contrast to the normal condition in which the air temperature decreases with altitude.

NTI: see *nocturnal temperature inversion*

plain: a region of generally uniform slope, comparatively level, having considerable extent, and not broken by marked elevations or depressions.

plateau: a comparatively flat area of great extent and elevation; specifically, an extensive land region considerably above the adjacent country or above sea level. Commonly plateaus are limited on at least one side by an abrupt descent, have flat or nearly smooth surfaces and are often dissected by deep valleys and surmounted by high hills or mountains, and have a large part of their total surface at or near the summit level.

refraction: the bending of light as it passes between two optical media that have different indices of optical refraction. See *Snell's Law* in this *Glossary*.

RM: ranch to market road (sometimes referred to as a farm to market road); symbolized on maps by a white rectangle with the road number inside

scientific notation: in scientific notation a number, *N*, is written like this —

$$N = a \times 10^{b}\text{, or } N = a\,E\,b\text{,}$$

where ***a*** is a decimal number and ***b*** is an integer. For example 412 is written as

$$N = 412 = 4.12 \times 10^{2}\text{, or } N = 4.12\text{ E02.}$$

The utility of the abbreviated form of number representation is illustrated by writing a number like N = 93,000,000,000. In scientific notation the trailing nine zeroes are not written. Instead, the number takes the form

$$N = 9.3\text{ E10.}$$

Snell's Law: a mathematical equation, first stated by Willebrord Snell in 1621, describing how light travels between two optically transparent media that are adjacent to each other, and share a flat interface between them. The arrangement of media, angles, light rays, and interface is shown in Figure 8-2. The equation

$$IOR_1 \times \cos(\theta_1) = IOR_2 \times \cos(\theta_2)$$

expresses the relation between the indices of refraction (IOR) and the angle of the light ray in each medium. Knowing any three of the four parameters allows the fourth parameter to be calculated.

A particular angle of interest is the ***Minimum Angle*** referenced in the discussion of Figure 8-3. It is the angle between the interface and the upper segment of ray ***c*** in Figure 8-3 when the lower segment of ray ***c*** is tangent to the interface between two layers of air.

The lower layer in Figure 8-3 has a smaller IOR than the upper layer does. The *Minimum Angle* for the ray comes directly from Snell's Law by substituting $\theta_2 = 0$. Then, one obtains

$$\cos(\theta_1) = IOR_2 / IOR_1$$

After substituting numerical values for the IORs, θ_1 can be calculated. In the Chapter 8 example, $IOR_1 = 1.00022395095$ and $IOR_2 = 1.00020428872$, yielding $\theta_1 = 0.359258$ degrees.

solid angle: see *Appendix*

steradian: see *Appendix*

sunset and **sunrise:** the times in the evening and morning when the upper edge of the disk of the sun is aligned with the horizon, which is unobstructed relative to the viewing location. It is assumed that the viewing location is in a level region on the earth's surface, and average atmospheric conditions are in force. For computational purposes this means that the center of the sun is geometrically 50 arc minutes below a horizontal plane. For an observer at sea level with a level, unobstructed horizon, under average atmospheric conditions, the upper limb of the sun will then appear to be tangent to the horizon. The 50 arc minute geometric depression of the sun's center used for the computations is obtained by adding the average apparent radius of the sun (16 arc minutes) to the average amount of atmospheric refraction at the horizon (34 arc minutes). (Source: United States Naval Observatory, http://www.usno.navy.mil/USNO/)

tinaja: a small pool of water in an earthen hollow or depression.

Trans–Pecos: the region of west Texas bounded by the southern border of New Mexico and the two rivers, Rio Grande and Pecos.

troposphere: the layer of air beginning at the earth's surface and extending upward 4–12 miles depending on the latitude. At the poles, the troposphere is only four miles thick, but it grows to a thickness of 12 miles at the equator.

true north: see *bearing.*

USGS: United States Geological Survey

—§—

Bibliographic Sources

Adams, Henry. *Practical Surveying and Elementary Geodesy.* New York: The MacMillan Company, 1913.

Bowditch, Nathaniel. *The New American Practical Navigator, 5th Edition.* New York: Edmond M. Blunt, 1821. [First edition published in 1802.]

Bowditch, Nathaniel. *The New American Practical Navigator, 16th Edition.* New York: E. & G. W. Blunt, 1846.

Bownds, Deric. "Marfa Stories told by Ruth Bownds Kershner", January 8, 1999. (Compiled by Susan Kershner Mundine, November 2001). Available from: Deric Bownds, http://www.dericbownds.net/Bownds_History/Marfa_Stories/MarfaStories.htm

Brueske, Judith M. "Letter to the Editor." *The Big Bend Gazette,* September 2005, Vol. 5, No. 9.

Brueske, Judith M. *The Marfa Lights – Being a collection of first-hand accounts by people who have seen the lights close-up or in unusual circumstances, and related material,* revised ed. Alpine, TX: Ocotillo Enterprises, 1989.

Bunnell, James. *Hunting Marfa Lights.* Benbrook, TX: Lacey Publishing Co., 2009.

Bunnell, James. *Marfa Lights.* DVD. Video presentation by James Bunnell. Cedar Creek, TX: Lacey Publishing Company, 2005.

Bunnell, James. *Night Orbs.* Cedar Creek, TX: Lacey Publishing Co., 2003.

Carter, W.T. M.W. Beck, and H.M. Smith (United States Department of Agriculture), and H.W. Hawker, E.H. Templin, and T.C. Reitch (Texas Agricultural Experiment Station), "Soil Survey (Reconnaissance) of the Trans–Pecos Area, Texas." *Bureau of Chemistry and Soils, United States Department of Agriculture, Series No. 1928, No. 35.* (Washington, D.C.: U.S. Government Printing Office, 1934).

Charnock, Richard. *Brad's Illustrated Hand-book to Spain and Portugal: A Complete Guide for Travellers in the Peninsula. With Maps, Town Plans, and Steel Illustrations* (London: W.J. Adams, 1867).

Clark, R.N. *Visual Astronomy of the Deep Sky.* Cambridge, England: Cambridge University Press and Sky Publishing, 1990.

Clements, C., C. Whiteman, and J. Horel. "Cold-Air-Pool Structure and Evolution in a Mountain Basin: Peter Sinks, Utah," *Journal of Applied Meteorology*, 42 (2002): 752.

"Connected. The Silver Spike and the Southern Pacific and Sunset Railroads." San Antonio *The Evening Light,* January 12, 1883, p. 1, col. 2.

Crawford, David L. "Photometry Terminology and Units in Lighting and Astronomy," *The Observatory* 117 (February, 1997): 14-18.

De Laborde, Alexander. *VIEW OF SPAIN; Comprising a Descriptive Itinerary, of Each Province, and a General Statistical Account of the Country; Including its Population, Agriculture, Manufactures, Commerce, and Finances; Its Government, Civil, and Ecclesiastical Establishments; The State of the Arts, Sciences, and Literature; Its Manners, Customs, Natural History, etc.* (London: Longman, Hurst, Rees, and Orme, 1809).

District Traffic Maps 1936-1990 [CD]. Austin, TX: Texas Department of Transportation, 2004.

Disturnell, J. *Mapa de los Estados Unidos de Mejico*, Revised Edition. New York, 1847. (Available from the Library of Congress)

Echols, Bevet Second Lieutenant William H. "Diary", Report of Secretary of War, in Message from the President of the United States to the Two Houses of Congress of the 36th Congress, 2nd Session, Senate Executive Document No. 1, Vol. II., Washington, D.C: 1860.

Edwards, Cindy. "Remembering the Early Days," *Update – Rio Grande Cooperative, Inc.*, May 2003.

"El Paso. Notes on Railroad Building—The Southern Pacific—Mexican-Central—Ice Factory, Special Telegram to The News." *The Galveston Daily News,* January 22, 1882, p. 1, col. 8.

Emory, Major William H. Boundary between the United States and Mexico. Map Number 2. Library of Congress Geography and Map

Division, Washington, DC 20540-4650 USA, 1855. Call number: G3701.F2 1855 .E6 TIL.

Emory, Major William H. *Report on the United States and Mexican Boundary Survey, Vol. I.* 34th Congress, 1st Session, Ex. Doc. No. 135. Washington, DC: 1857.

Fletcher, Henry T. "Marfa," published in "Texas Collection," edited by H. Bailey Carroll, *Southwestern History Quarterly* 48, no. 2 (1944): 295.

"Foreign Intelligence. Railroads Interrupted by Floods." *Galveston Daily News*, November 2, 1879, p.1, col. 6.

Froebel, Julius. *Seven Years' Travel in Central America, Northern Mexico, and the Far West of the United States.* London: Richard Bentley, 1859.

General Electric Basic A19 Product Code 41026. Available from: http://genet.gelighting.com/LightProducts/Dispatcher?REQUEST=CONSUMERSPECPAGE&PRODUCTCODE=41026

Greene, Nathaniel R. "Shedding Light on the Candela", *The Physics Teacher* 41 (October 2003): 409-414.

Gregg, John Ernest. "The History of Presidio County." MA thesis, University of Texas, Austin, TX, August 1933. [Excerpts reprinted in *Voice of the Mexican Border,* Centennial Edition, 1936.]

Hall, Michael. "The Truth is Out There." *Texas Monthly*, June 2006.

Harding, C. Francis, and A.N. Topping, "Headlight Tests," *Transactions of the American Institute of Electrical Engineers* 33 (May 17-December 31, 1910): 1053-1081.

Hartz, Second Lieutenant Edward L. "Diary", Report of Secretary of War, in Message from the President of the United States to the Two Houses of Congress of the 36th Congress, Executive Documents of the United States, 1st Session, Vol. II, No. 2, Part 2, Washington, D.C.: 1859.

Hodgson, James. *A System of the Mathematics, Printed for Thomas Page.* London: William and Fisher Mount, 1723.

"Intensity of Lights," in *Code of Federal Regulations, Title 33, Subchapter E, Paragraph 84.15.* Available from: http://ecfr.gpoaccess.gov/cgi/t/text/text-idx?c=ecfr&sid=a6396ba406dff0936af39579b8b54acd&rgn=di

v8&view=text&node=33:1.0.1.5.39.0.26.8&idno=33

Kloppers, Johannes Christiaan. "A Critical Evaluation and Refinement of the Performance Prediction of Wet-Cooling Towers." Ph.D. dissertation, University of Stellenbosch, Stellenbosch, South Africa, December 2003.

Lathan, William, "Account of a Singular Instance of Atmospheric Refraction," *Philosophical Transactions of the Royal Society of London*, 88, (1798): 357-360.

Lettau, Heinz and Ben Davidson, (eds.), *Exploring the Atmosphere's First Mile* London: Pergamon Press, 1957.

Lighting and Illumination Committee. "State Regulation of Motor Car Headlamps." *Journal of the American Institute of Electrical Engineers* 41, no. 3 (March 1922): 223-225.

MacDonald, William, ed. *Select Documents Illustrative of the History of the United States 1776-1861.* New York: MacMillan, 1898. [Available from Google Books.]

Malm, W.C., D.E. Day, S.M. Kreidenweis, J.L. Collett, and T. Lee, "Humidity-dependent Optical Properties of Fine Particles During the Big Bend Regional Aerosol and Visibility Observational Study." *Journal of Geophysical Research* 108, D9 (2003): 4,279.

McAnulty, William N., "Geology of Cathedral Mountain Quadrangle, Brewster County, Texas", *Bulletin of the Geological Society of America* 66, no. 5 (May 1955): 531-578.

Middleton, W.E. Knowles. *Vision Through the Atmosphere.* Toronto: University of Toronto Press, 1952.

Miles, Elton. *Tales of the Big Bend.* College Station, TX: Texas A&M University Press, 1976.

Mills, Evans. *Technical and Economic Performance Analysis of Kerosene Lamps and Alternative Approaches to Illumination in Developing Countries.* Berkeley, CA: Lawrence Berkeley National Laboratory, 2003.

Mills, Evan. "The Specter of Fuel-Based Lighting", *Science* 308 (2005): 1263-1264.

Moore, David W. "Headlight History and Harmonization," *Report No. UMTRI-98-21*. Ann Arbor, MI: The University of Michigan Transportation Research Institute, June, 1998.

Moran, Paul. "The Mystery of the Texas Ghost Light." *Coronet*, July 1957.

"Motor Headlights Must be Tested," *The New York Times*, May 12, 1918, Section 5, p. 3, col 1, Sunday edition.

NASA, "International Satellite Cloud Climatology Project (ISCCP)," http://isccp.giss.nasa.gov/.

NOAA Earth System Research Laboratory, Physical Sciences Division, "3-Hourly NCEP North American Regional Reanalysis (NARR) Composites," NOAA. Available from: http://www.esrl.noaa.gov/psd//cgi-bin/data/narr/plothour.pl

NOAA, "NOAA Celebrates 200 Years of Science, Service, and Stewardship," National Oceanic and Atmospheric Administration. Available from: http://celebrating200years.noaa.gov/magazine/signals/welcome.html

Peacock, George, ed. *Miscellaneous Works of the Late Thomas Young, Volume II.* London: John Murray, 1855).

Phillips, Sir Richard. *A Million of Facts, of Correct Data, and Elementary Constants, in the Entire Circle of the Sciences, and on all Subjects of Speculation and Practice.* London: Darton & Co., 1848.

Pidgeon, D. "A Green Sun," *Nature*, 37, March 8, 1888, 440.

Pitchford, Marc L. and others. *Big Bend Regional Aerosol and Visibility and Observational Study [BRAVO], Final Report—September 2004*. Available from: http://vista.cira.colostate.edu/improve/Studies/BRAVO/Studybravo.htm

Powell, A. Michael. *Trees and Shrubs of the Trans-Pecos and Adjacent Areas.* Austin, TX: University of Texas Press, 1998.

"Presidio," *The Galveston Daily News*, December 17, 1882, p. 7, col. 5.

Raht, Carlysle. *The Romance of Davis Mountains and Big Bend Country,* Texana Edition. Odessa, TX: The Rahtbooks Company, 1963.

Rebert, Paula. *La Gran Línea.* Austin, TX: University of Texas Press, 2001.

Reed, S.G. *A History of the Texas Railroads.* Houston, TX: St. Clair Publishing Co., 1941.

Rodriguez, Jose Policarpo. *The Old Guide.* Dallas, TX: Publishing House of the Methodist Episcopal Church, [n.d.]. (Dictated to Rev. D.W. Carter, D.D., from 1892 to 1897.)

Santleben, August. *A Texas Pioneer.* New York: The Neale Publishing Company, 1910.

Smith, C. Michie. "Observations on a Green Sun, and Associated Phenomena," *Nature,* 30, August 7, 1884, 347. (Abstract of a paper read before the Royal Society of Edinburg, July 7, 1884.)

Smith, Thomas, and John Guild, "The C.I.E. colorimetric standards and their use." *Transactions of the Optical Society* 33, no. 3 (1931): 73-134.

Smithers, W.D. *Chronicles of the Big Bend.* Austin, TX: Texas State Historical Association, 1999.

Society of Physics Students. "An Experimental Analysis of the Marfa Lights," Dallas, TX: University of Texas at Dallas, December 10, 2005. Available from: http://www.spsnational.org/wormhole/utd_sps_report.pdf [The link in the PDF to the videos they produced no longer works.]

"Southern Pacific Company." In *The Rand-McNally Official Railway Guide, and Handbook, for the United States and the Dominion of Canada,* no 3, September 1886, 294. Chicago, IL: National Railway Publication Company, 1886.

Southern Pacific Lines, The Galveston, Harrisburg & San Antonio Railway, Brief History of El Paso Division 1881-1925, originated by Clifford Morrill and compiled by Hugh B. Wilson. El Paso, TX: [n.p.], 1925. (Issued April 30, 1925.)

Stone, Jack A. and Jay H. Zimmerman, "Index of Refraction of Air," in Engineering Metrology Toolbox, Physical Measurement Laboratory, NIST, http://emtoolbox.nist.gov/Wavelength/Documentation.asp

Sylvania "OSRAM Sylvania Products, Part Number H6054." Available from: http://www.sylvania.com/en-us/products/Pages/default.aspx

Symons, G.J., F.R.S., editor. "The Eruption of Krakatoa, and Subsequent Phenomena," *Report of the Krakatoa Committee of the Royal Society, 57 and 59.* Ludgate Hill, London: Trubner and

Company, 1988.

Thompson, Cecilia. *History of Marfa and Presidio County, Texas 1535-1946, Volume 1, 1535-1900.* Marfa, TX: The Presidio County Historical Commission, 1985.

Thompson, Cecilia. *History of Marfa and Presidio County, Texas 1535-1946, Volume 2, 1901-1946.* Marfa, TX: The Presidio County Historical Commission, 1985.

Thomson, Sir William [Lord Kelvin]. "Electrical Units of Measurements." Lecture delivered at the Institution of Civil Engineers in London on May 3, 1883. Reprinted in Thomson, Sir William. *Popular Lectures and Addresses, Volume 1: Constitution of Matter.* London: MacMillan and Company, 1891.

"Through to California." *Galveston Daily News.* January 24, 1883, p. 7, col. 3.

US Geographical Survey (USGS) Website http://www.usgs.gov/pubprod/maps.html

USGS Map Product 301227, "Alpine, TX Historical Map Geopdf 30x60 Grid 125000-scale 1895." Available from: http://store.usgs.gov/b2c_usgs/catalog/query/(ctype=areaDetails&xcm=r3standardpitrex_prd&carea=%24ROOT&layout=6_1_61_48&uiarea=2)/.do?query=301227&lastVisited=areaDetails

Warranty deed. Grantee: FW Jordan; Grantor: Mitchell, FA et ux; Transaction date: 06-01-1929; Record date: 03-01-1930. *Deed Book 87.* Marfa, TX: Presidio County Clerk's Office, pp. 162-163.

Westinghouse HF100R, Product Code 3740300. Available from: http://www.westinghouselighting.com/light-bulbs/hid-bulbs/mercury-vapor/100-watt-ed23-1-2-hid-mercury-vapor-light-bulb-3740300.aspx

Whiting, W.H.C. "Whiting Diary," *Publications of the Southern History Association* 10, no. 2 (March, 1906).

Williams, Judge O.W. "Alsate – The Last of the Chisos Apaches." *Frontier Times,* February 1944, Vol. 21, No. 5.

Williams, Oscar Waldo. "Alsate." in *By the Camp-fire in the Southwest,* 8-13. Dallas, TX: Martin Stationery Co., 1934.

Williams, O.W. *Alsate, The Last of the Chisos Apaches.* [n.p., n.d.]. [The version of the pamphlet the authors consulted is among the holdings of Southern Methodist University's DeGolyer Library in Dallas, TX; the call number is E99 A6 W55.]

Williams, O.W. *Pioneer Surveyor–Frontier Lawyer, The Personal Narrative of O. W. Williams, 1877-1902*, ed. S.D. Myres, 222-249. El Paso, TX: Texas Western Press, The University of Texas at El Paso, 1968.

Wilson, R. "The Blue Sun of 1950 September." *Monthly Notices of the Royal Astronomical Society,* 111, (1951): 477.

—§—

About the Authors

ROBERT WAGERS holds a Ph.D. degree in electrical engineering from Stanford University. He spent most of his professional career developing technology for optical and electrical signal processing systems.

Two of his research activities are specifically relevant to the subjects investigated in this book. His three years of postgraduate research in plasma physics at the Institute for Plasma Research at Stanford University position him well to assess the common assertion that "ball lightning" is responsible for the Marfa lights. Similarly, sixteen years of research on ultrasonic wave phenomena in piezoelectric crystals enable a critical evaluation of the assertion that the Marfa lights come from crystals under pressure. Dr. Wagers has technological experience in two other areas that are particularly relevant to this book. He spent two years developing imaging systems that utilize charge-coupled devices, and he served as technical leader in the design of a hyperspectral imaging satellite for the U.S. Air Force.

Dr. Wagers is a Rhodes Scholar and Fellow of the Institute

of Electrical and Electronics Engineers. He has 56 critically-reviewed scientific publications and 12 U.S. Patents. He is retired and conducts investigations, such as the ones in this book, for the pure joy of discovery.

—§—

JUDY WAGERS has master's degrees in both English and business administration. She began her professional career as an editor and project manager for a California publisher of children's educational materials.

Subsequently, Judy joined the Department of Obstetrics and Gynecology at the University of Texas Southwestern Medical Center in Dallas, Texas, where she worked for the next 36 years. As department manager, her primary responsibilities involved oversight of all business operations for this department of 300 faculty, staff, residents, and fellows. Since her retirement, she has been writing about the history of the department for the UT Southwestern website.

Intrigued by mystery and adventure, Judy was eager to participate in the field trips and historical research needed to understand the mysterious lights. As the illustrator and editor-in-chief of this twosome, she was charged with the challenge of translating the technical subjects into lay terms in a way that will capture the reader's interest.

—§—

Made in United States
North Haven, CT
10 November 2026

22095618R00143